THE BEHAVIOUR OF THE DOMESTIC CAT

John W.S. Bradshaw

Department of Biology
University of Southampton
Southampton SO9 3TU
UK

CABI *Publishing*

CABI is a trading name of CAB International

CABI Head Office
Nosworthy Way
Wallingford
Oxfordshire OX10 8DE
UK

CABI North American Office
875 Massachusetts Avenue
7th Floor
Cambridge, MA 02139
USA

Tel: +44 (0)1491 832111
Fax: +44 (0)1491 833508
Email: cabi@cabi.org
Web site: www.cabi.org

Tel: +1 617 395 4056
Fax: +1 617 354 6875
Email: cabi-nao@cabi.org

A catalogue record for this book is available from the British Library, London, UK.

ISBN-13: 978-0-85198-715-6
ISBN-10: 0- 85198-715-X

First published 1992
Reprinted 1998, 2000, 2002
Transferred to print on demand 2007

Printed and bound in the UK by CPI Antony Rowe, Eastbourne.

THE BEHAVIOUR OF THE DOMESTIC CAT

Contents

Preface

The behaviour of the domestic cat, *Felis catus*, has many features that set it apart from other common domestic animals, such as cows, sheep, pigs, horses, and even its fellow carnivore, the dog. Cats seem to have effected a unique and successful compromise between reliance on man and the retention of behaviour patterns that permit an independent existence. And yet, several recent surveys have shown that the cat is the preferred pet of the most 'modern' type of pet owner, valued for its combination of an affectionate nature with a degree of independence, as well as more prosaic qualities such as cleanliness and convenience. The cat may well turn out to be the dominant pet species in the early part of the 21st century, yet until recently many aspects of its behaviour have been poorly documented. Our increasing knowledge of the behaviour patterns that make up the very special character of the cat can be made use of in several ways: to improve the treatment of behaviour that we perceive as abnormal, or at least undesirable; to provide a point of comparison for other, undomesticated felids; to enhance the cat owner's appreciation of their pet; and so on. This book brings together many disparate studies of the behaviour of *Felis catus*, and attempts to marry the more mechanistic approach (how do patterns of behaviour come about?), with the functional (what are those patterns for, and how do they benefit the animal?). It may seem strange to ask functional questions about the behaviour of a domestic animal, whose feeding and breeding are both largely controlled by man, but many cat biologists have done so, and successfully at that.

From the biological perspective, it is important to know from the outset how any one species fits in with other species, in terms of its genetic relatedness, and the characteristics that make that species unique. The first chapter deals with these issues: the relationship between the domestic cat and the rest of its family, the Felidae, and the basis for its domestication. The next two chapters attempt to describe the world in the cat's own terms: what it can see, hear, smell and feel, and how it can relate all this incoming information together so that its surroundings make sense. This is worth establishing at the outset, because their world does not have the same appearance as ours. Furthermore, cat intelligence has been portrayed in many ways, from the early behaviourists' picture of a near-automation, to the anthropomorphic sentimentalizations of the popular literature; Chapter 3 redresses the balance in favour of modern scientific thinking.

Subsequent chapters deal with more specialized aspects of cat behaviour, such as reproduction, development, communication, hunting and feeding. Chapters 8, 9 and 10 are based around a common theme, and one that could not have been used if this book had been written 15 or so years ago, for it is only during that time that the cat's social abilities have become well-known. The complex social interactions between cats, many of which we do not yet fully understand, have forced biologists to reinterpret not only the cat's relationship with its conspecifics, but also the bond with its owner, and considerations of its welfare.

To the veterinary surgeon, none of this information will be any more than interesting without practical advice on dealing with behavioural problems presented to them by clients. The concluding chapter, written by Peter Neville, contains a wealth of practical advice drawn from his extensive 'hands-on' experience of treating such problems, in the clinics he has run at veterinary colleges and hospitals.

For those interested in reading further, two recent books can be regarded as complementary to this one. *The Natural History of the Wild Cats*, by Andrew Kitchener, extends the story to the other members of the Felidae. More comprehensive treatments of many of the topics are contained in the multiauthor book edited by Dennis Turner and Patrick Bateson, *The Domestic Cat: the biology of its behaviour.* Further suggestions can be gleaned from the list of references, which has been deliberately kept short, by using review articles rather than research papers wherever possible, to avoid breaking up the text with large numbers of citations.

Many of my friends and colleagues have helped me with this book, by making suggestions, providing references, allowing me to present their unpublished research, and reading drafts of chapters. In no particular order, they are Michael Mendl, Sandra McCune, Gillian Kerby, Warner Passanisi, Chris Thorne, Ian Robinson, Helen Nott, David Macdonald, Rory Putman, Sarah Brown, Sarah Lowe, Debby Smith, Katy Durman, Stuart Church and Steve Wickens. The responsibility for what appears on the printed pages is, of course, mine alone. My sincere gratitude must also go to my family for tolerating my year as a recluse, to Lucy for producing the kittens figured in Chapter 4, and to Michael Toms for drawing them and producing the cover illustration and many of the other figures in this book (indicated by his initials after the caption). The observations of rescued cats described in Chapter 10 were carried out at St Francis Animal Shelter, Horton Heath, and I am grateful to the Warden, Mrs Anne Hillman, and her family, for their tolerance of us over the past four years. My own research, and that of my students, reported here has been supported by the Waltham Centre for Pet Nutrition, the Science and Engineering Research Council, the Universities Federation for Animal Welfare, the Cats Protection League, and the University of Southampton, to all of whom I express my gratitude.

The Cat:
Domestication and Biology

The domestic cat is by far the most numerous of all the cat family, the Felidae, many of whose members are listed as approaching or vulnerable to extinction (Kitchener, 1991). While a detailed discussion of the derivation of the domestic cat from wild cats is outside the scope of this book, and is in any case far from being completely understood, from a biological point of view such relationships are crucial to a full understanding of behaviour. This is because one of the basic principles of modern ethology is that species-specific behaviour patterns contain an inherited component, and therefore can be compared from species to species as if they were a morphological character, like the shape of the skull. While it is possible to study the behaviour of the domestic cat in isolation, we are likely to gain more insight into the origins of many behaviour patterns if we can compare them to those of related undomesticated species. Having said that, behavioural studies of closely related species are sparse at present, and the most detailed comparisons can only be made with the 'big cats', notably the lion.

Five of the ten species of big cat are grouped in the genus *Panthera*, including the lion, tiger and leopard. This genus is characterized by its ability to roar, and inability to purr, due to modifications to the hyoid bone at the base of the tongue. The small cats are now all classified into the genus *Felis* (Kitchener, 1991), divided into about 25 species (Table 1.1). They share a number of behavioural differences from the large cats, including the ability to purr continuously, and a tendency to feed in the crouched position, rather than lying down. Many of the species of *Felis*, and their various races, are not well studied, although the genus contains some well-known types, such as the puma, wild cat and ocelot.

The species thought to be the main ancestor of the domestic cat is the Wild Cat *Felis silvestris* (Clutton-Brock, 1987). Previously divided into

Table 1.1. The members of the genus *Felis*, indicating one of their common names, and their approximate ranges and habitat preferences (derived from Kitchener, 1991). One group probably originated in the New World, and is now largely confined to South America. The other two groups have Old World origins; the puma and jaguarundi are thought to have invaded the Americas via the Bering land bridge. The only subspecies listed are those of *F. silvestris*, of which the domestic cat is one.

Common name	Species	Range: habitat
Ocelot group – New World small cats		
Ocelot	*pardalis*	Arizona to Uruguay: wide range of habitats
Margay	*wiedii*	Costa Rica to N Argentina: mainly forest
Tiger Cat	*tigrina*	Costa Rica to N Argentina: forest
Geoffroy's Cat	*geoffroyi*	Bolivia to Patagonia: woodland, bush
Kodkod	*guigna*	Chile: forest
Pampas Cat	*colocolo*	Ecuador to Patagonia: wide range of habitats
Mountain Cat	*jacobita*	Andes: steppe
Pantherine group – Old World small cats		
Leopard Cat	*bengalensis*	S and E Asia: woodland, forest, scrub
Iriomote Cat	*iriomotensis*	Iriomotejima Is. (Japan): rainforest
Rusty-spotted Cat	*rubiginosa*	S India, Sri Lanka: wide range of habitats
Fishing Cat	*viverrina*	S and SE Asia: near water
Flat-headed Cat	*planiceps*	Borneo, Sumatra, Malaya: forest near water
African Golden Cat	*aurata*	Senegal to Zaire, Kenya: high rain forest
Asian Golden Cat	*temminckii*	Himalayas to SE Asia: high forest
Bay Cat	*badia*	Borneo: tropical forest and rocky scrub
Serval	*serval*	Africa: mainly savannah
Caracal	*caracal*	India to S Africa: wide range of habitats
Puma	*concolor*	S Canada to Patagonia: forest, swamp, bush
Jaguarundi	*yagouaroundi*	Arizona to N Argentina: lowland forest
Domestic cat group – Old World small cats		
Chinese Desert Cat	*bieti*	N China, Mongolia: steppe and mountain
Jungle Cat	*chaus*	Egypt to India: wide range of habitats
Sand Cat	*margarita*	Sahara to Turkestan: desert and semi-desert
Black-footed Cat	*nigripes*	Southern Africa: desert, grassland
Pallas' Cat	*manul*	Iran to W China: steppe, desert
Wild cat	*silvestris*	
European Wild Cat	*s. silvestris*	Scotland to Georgia (USSR): mainly forest
African Wild Cat	*s. lybica*	Africa: wide range of habitats
Indian Desert Cat	*s. ornata*	SW Asia, N India: semidesert, steppe
Domestic Cat	*s. catus*	Worldwide: association with man

several species and races, *F. silvestris* is now thought to be a single species that varies continuously from the north-west to the south-east of its range, between the Scottish Wild Cat *F. s. silvestris*, the African or Arabian Wild Cat *F. s. lybica*, and the Indian Desert Cat *F. s. ornata*. The northern forms are the more thick set, with heavier fur, while in the tropics the same species is much more fine-limbed and light-coated, to suit the climate. There are also forms between these two extremes, and it is some of these that most closely resemble the domestic cat, which is now formally classified as *F. s. catus*. The geographical variations in the appearance of *F. silvestris* must presumably also be matched by differences in social behaviour, for while the domestic cat superficially resembles the European, the latter is impossible to domesticate. The *lybica* form is thought to be much more adaptable to living alongside and within human settlements, and is therefore likely to have been the ancestor of *catus*. Other evidence to support this origin comes from archaeology (see below), alloenzyme studies (Randi and Ragni, 1991), and the names used to describe cats, many of which, including 'puss', 'tabby' and the word 'cat' itself, have North African or Middle Eastern origins. It has also been suggested that the oriental breeds of cat were originally derived from the light-bodied Indian Desert Cat.

From time to time it has been suggested that other species of *Felis*, whose ranges overlap with *lybica*, might also have contributed to the modern *catus*, including Pallas' Cat *F. manul* and the Jungle Cat *F. chaus*. The former has recently been ruled out, because it has a different number of chromosomes than *F. silvestris* (see Wayne *et al.*, 1989). Remains of the latter have been found in Egypt alongside those of *silvestris*, and remains of cats with skulls intermediate in size between *chaus* and *lybica* have been found in Egyptian tombs. *F. chaus* can indeed form fertile hybrids with *F. silvestris*, so it is possible that *chaus* has made a small contribution to the genetics of the modern domestic cat.

The colour forms of *F. catus* (excluding the 'fancy' breeds) have arisen by somatic mutation from the original striped tabby coat of *lybica*. The main variations are the blotched tabby, which has three black stripes on the back and spiral or circular stripes on the flanks, the sex-linked orange tabby (which produces orange males and females, and the heterozygous tortoiseshell and calico females), the plain black, and the piebald white, with both the pure forms and many combinations evident in the modern domestic population. Most of the specific breeds, such as the Persian, Abyssinian and Turkish Van, have probably been produced by selective breeding of *catus*, rather than by cross-breeding with other species of *Felis*. Part of this process has been a selection for breed-specific behavioural traits, which will be discussed below.

The population genetics of each of the major coat-colour genes have been studied in some detail. While these genes are unlikely to affect behaviour directly, it has been shown in other carnivores, such as mink and foxes, that coat colours can be associated with heritable tendencies towards fear and aggression, so it is worthwhile considering the distributions of coat genes in present-day populations. The perpetuation of many highly distinctive coat genotypes in populations of domestic animals appears to be a product of man's preference for novelty; in wild animals a new mutation is often lost by chance, through genetic drift, or, if highly advantageous, displaces its original form and becomes fixated in the population. In the domestic cat, the effects of both artificial and natural selection can be seen. For example, the blotched tabby allele seems gradually to be displacing the wild-type striped tabby allele, even in feral populations, although the advantage that it confers has not been identified. This allele probably originated in Britain, some time before the 17th century, and has been

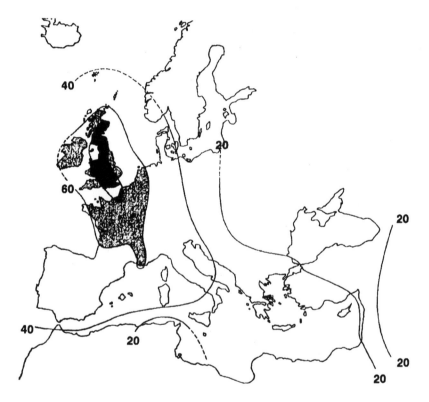

Fig. 1.1. The distribution of the blotched-tabby allele in Europe and around the Mediterranean basin. Areas with over 80% of cats carrying this allele are given solid shading, those with over 60% are stippled, and the 40% and 20% contour lines are indicated by the numbers. This allele is also common in areas first colonized from Britain, including New Zealand, Canada, Australia and New Zealand. (Redrawn from Todd, 1977.)

spread by colonization, particularly to Australia and North America (Todd, 1977). A second, more recent origin in or to the east of northern Iran has also been detected (Fig. 1.1). By contrast, the dominant white phenotype, although often preferred by people, confers disadvantages such as deafness and an increased susceptibility to skin cancers, which presumably act to suppress its frequency in natural populations.

In an urban population, Clark (1975) found that human preference had little effect on the selection of coat colours. While pet populations generally had lighter coats than did strays (at up to six separate genetic loci), the majority of the pets were neutered, and therefore the effect of human preference for colour was unlikely to be transmitted to the next generation. However, the global distribution of particular genes seems to be highly influenced by human activities, and particularly by the long-standing custom

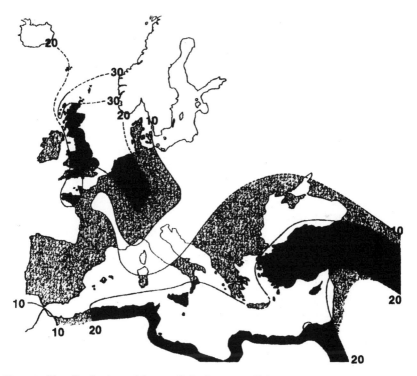

Fig. 1.2. The distribution of the sex-linked orange allele in Europe and around the Mediterranean basin. Areas with over 20% of cats carrying this allele are given solid shading, and those with over 10% are stippled. This allele can produce tortoiseshell, torbie, calico or marmalade coats, depending on the sex of the cat and the other genes it occurs with. This mutation probably first gained a foothold in Asia Minor, and spread via sea routes and waterways to north-western Europe, where it has become particularly favoured in the Scottish Isles. (Redrawn from Todd, 1977.)

of carrying cats on cargo boats. It appears that cats selected for this purpose tended to be of unusual types that might have disappeared had they remained part of a larger population. The sex-linked orange allele seems to have been affected in this way; it probably first became established in Asia Minor, and was then transported through the trading routes of the Mediterranean and, via the Seine and Rhone valleys, to London. However, existing large cat populations resulted in its only achieving a low frequency in these latter areas (Fig. 1.2). Higher frequencies in Scotland and in North Africa appear to be a result of human preference for orange and/or tortoiseshell in those areas.

Domestication

Juliet Clutton-Brock (1987) describes the cat as 'an exploiting captive' and a 'carnivore that enjoys the company of man'. The cat is neither a man-made species like the dog, nor simply an animal made captive for utilitarian purposes, like the elephant. The history of its domestication is therefore unusual, and is also comparatively recent. Moreover, because there has been so little change in form from *lybica*, particularly in the skeleton, the archaeological evidence is not conclusive. The process of domestication probably started in Egypt in about 4000 BC, since the remains of *Felis silvestris*, and other cats such as *F. chaus*, are found in Egyptian tombs of that period. However, the remains of other carnivores have also been found at prehistoric sites, and it is unclear whether these earliest remains are of domestic cats, or of animals killed for their pelts.

Egyptian paintings and sculptures dating from about 1600 BC onwards give conclusive evidence of domestic status, as cats are depicted sharing many of man's activities, such as eating and hunting (Serpell, 1988). Theories of the initial function of domestication are divided between the utilitarian and the cosmic. The economy of Egypt was, at that time, largely based on grain, and so the ability of cats to control outbreaks of granivorous rodents might have stimulated at least a measure of encouragement from the populace, perhaps in the form of supplementary feeding and the provision of protected nest sites (the latter in view of the possible infanticidal tendencies of male cats, discussed in Chapter 8). It has even been suggested that, under these circumstances, cats initially became commensal, and then started the process of domestication themselves. Whatever the initial route, cats came to play a great part in the Egyptian pantheistic religions, culminating in the elevation of the cat goddess Bastet into the national deity in about 950 BC. The sacred status of cats led to their being mummified in vast numbers, leading to the bizarre use of many tons of these corpses as fertilizer, following their excavation in the early part of our own century.

Table 1.2. Numbers of cats in selected countries for which figures are available.

Country	Cats (millions)	Cats per human
Austria	12.3	0.879
New Zealand	1.2	0.383
Algeria	7.0	0.357
Denmark	1.6	0.303
Israel	1.0	0.253
USA	54.0	0.231
Indonesia	30.0	0.200
South Africa	6.0	0.200
Poland	6.4	0.178
France	8.0	0.148
Australia	1.1	0.146
Ireland	0.5	0.145
Canada	3.4	0.140
Switzerland	0.9	0.140
Romania	3.0	0.134
Venezuela	1.6	0.112
Hungary	1.2	0.112
Netherlands	1.5	0.105
Finland	0.5	0.104
Belgium	1.0	0.101
UK	5.3	0.090
Italy	4.6	0.080
Nigeria	6.0	0.075
Iceland	0.02	0.074
Chile	0.8	0.071
Colombia	1.5	0.055
West Germany	3.2	0.052
Sweden	0.3	0.036
Senegal	0.2	0.026
Brazil	2.5	0.021
Niger	0.01	0.018
Japan	2.0	0.017
Burundi	0.02	0.004
India	1.7	0.002
Taiwan	0.03	0.002
Total	171	
Average		0.082

Based on data in Anon., 1990.

The spread of the domestic cat from Egypt appears to have taken place rather slowly, indeed their religious status seems to have hindered their export as pets or as controllers of rats and mice. The Romans appear to have favoured the much less tractable ferret and polecat for the latter purpose, until about 400 AD, when they began to adopt the cat. Domestic cats had reached India by about 200 BC, and thereafter spread to the Far East. In modern times their distribution has become global, but with marked differences between countries (Table 1.2). Cats are evidently most popular in the Antipodes, North America and Western Europe, although there are some cat-loving countries outside these areas, notably Algeria, Israel and Indonesia. The extent to which these cats are 'owned' will vary considerably from one country to another; for example, many Mediterranean cities contain large populations of strays. The extent to which the stray population is included in the figures in Table 1.2 will vary from country to country, and this variation may account for some of the apparent discrepancies between neighbouring countries.

In the USA there has been a recent trend for cat populations to increase, apparently at the expense of dogs. For example, in 1983 the US dog population was estimated at 55.6 million, with 42.5% of households owning at least one dog, but by 1987 the population had dropped to 52.4 million, in 38.2% of households. In 1983 cats, at 52.2 million, were owned by 28.4% of households; by 1987 they had overtaken dogs in numbers (54.6 million) if not in the number of households (30.5%). This trend seems likely to be repeated in Europe, giving substance to the idea that the cat is the pet of the future.

In behavioural terms, domestication has probably had less effect on the cat than on any other domestic mammal. The changes that have taken place seem to be of three kinds:

1. reduction in brain size, often correlated in other domesticated animals with a reduced sensitivity to uncongenial stimuli;
2. modification of the hormone balance, mainly by reduction in size of the adrenals;
3. neotony, the persistence of some juvenile behaviour characters into the adult.

Certain of these changes may have become necessary as the population density of cats increased, as they moved from agricultural communities into towns. However, the skulls of cats placed in Egyptian tombs at a time when the cat was thought to have been at least partially domesticated, indicate that at that stage the brain was at least as large as that of modern *lybica*, and distinctly larger than those of modern *catus*. The stable and organized social structure evident in groups of feral cats, genetically almost identical to house cats, demonstrates that association with man has little altered the cat's wild behaviour patterns. It is this quality that has made the domestic cat a suitable object of study for the modern sociobiologist, from whose work many of the complexities of cat sociality are beginning to emerge. Patterns of predatory behaviour are practised from an early age by kittens, and are used to considerable effect by adults; in most breeds of dog the predatory sequence is incomplete. Neotony is much less apparent in cats than in dogs, although such behaviour patterns as purring and treading with the forepaws, both characteristic of kittens, may be more commonly expressed in adult domestic cats than in their wild counterparts.

Man's incomplete control of the sexual behaviour of cats may help to perpetuate this wild character. Increasingly, man controls the breeding of most cats through neutering (depending on the area, 60–90% of pet cats in the UK are neutered). However, those that are not neutered are usually allowed to select their own mates, perpetuating the mating system. The effects of this can be seen in differences in the frequency of coat colours between areas where cats are allowed to breed freely, and areas where most individuals are neutered.

Breeding is much more tightly controlled in the 'fancy' breeds, where mates are selected on the basis of the predicted outcome of particular combinations of morphological characters, although chance matings can result in the transfer of 'fancy' genes into the general population. In contrast to the situation in most domesticated animals, including the dog, specific breeds make up a small minority of the cat population – as little as 6–7% in both the UK and the USA. There has been little direct study of the behavioural characteristics of even the commonest breeds, but the experience of cat-show judges and breeders can be used to produce overall descriptions (Hart, 1979), although all are agreed that differences between individual cats can be as great as differences between breeds. Persians tend to be less active than other breeds; their long coats may make sitting on a lap for any length of time uncomfortable, possibly the reason why they seem to prefer to be stroked when sitting on the floor or on a chair. Abyssinians, which are short haired, also tend not to be lap-cats. The extent to which such behavioural traits are inherited has not been determined, but they could be explained simply by changes in certain stimulus thresholds.

An extreme example of this can be found in the poor maternal behaviour of many blue-eyed white cats. The combination of genes that brings about these colours also often results in deafness, which could be thought of as an extreme raising of auditory thresholds. Such queen cats are unable to hear the cries of their kittens, and are therefore less responsive to them. A further behavioural impairment is found in some Siamese cats, where the eyes and brain are in some way wrongly connected, resulting in impaired vision and, in some affected individuals, a compensatory squint.

Those with the slender oriental body type, such as the Siamese and Oriental Shorthairs, do show distinct behavioural differences from breeds with the stockier 'cobby' or occidental body type, such as the British Short-hair and the Persian. The oriental body type was first reliably described in Siam in the 18th century, but may well have originated elsewhere, at an earlier date. The average modern example of this type is, by comparison with typical cross-bred cobby cats, noisy, active, highly responsive to external stimuli and more trainable, suggesting a higher capacity to learn, and/or a stronger social bond to the owner. Many Siamese cats actively seek out affection from people, although some authorities attribute this to a desire for warmth because of their thin coats. The Burmese types usually display the Siamese characteristics, but to a lesser degree, and in particular are less vocal. There does appear to be a discontinuity between the oriental and occidental breeds, as reflected in their behaviour, which may lend credence to the idea that they are derived from different races or sub-species of *F. silvestris*.

Relationship with man

Cats have a unique degree of flexibility in their dependence on man. In the UK, something between one and two million cats do not have specific owners, but rely on hunting, scavenging for waste food, and handouts from cat lovers. However, it is unlikely that this population could be sustained at any substantial level without man's support; apart from the supply of food, in a temperate climate the survival of kittens is very poor except in the warm, dry conditions provided by human habitations. In warmer countries there is often even less of a distinction between the house cat, the stray, the partially reliant, and the truly independent feral, and in some areas, such as the Middle East, there appears to be little distinction between feral individuals of the domestic cat, and wild *F. silvestris*, raising the possibility that extensive interbreeding with the wild ancestor has occurred.

Perhaps the cat has needed this flexibility to survive the changes in the relationship between man and his cat that have taken place over the centuries, veering from worship to revulsion. In contrast with the deification of the domestic cat by the Egyptians, Christianity adopted the cat as an almost universal scapegoat, particularly from the 13th century onwards. Cats, particularly black ones, were regarded as a product of the Devil. Their partly nocturnal habits, and the blood-curdling cries they make during fights, may have helped to single them out in this regard. Similar associations with evil are common in Oriental folklore. This attitude persisted, at least in English-speaking countries, until the beginning of this century, and there is still a substantial proportion of the public that expresses a dislike of cats. The reasons for this are obscure, but may be due to a combination of the old superstitions with distrust stemming from personal experience of the ambivalent relationship between cat and man.

Leaving such prejudices aside, the relationship between cat and man can be described as symbiotic and mutually beneficial. The cat gains a warm shelter, and a reliable supply of food, both essential for the raising of healthy kittens. The owner may assist the cat in defending the core area of its territory, by discouraging other cats from stealing food provided for their own cat. Some of the cat's psychological needs may also be met, through social interactions with the owner, such as play, and displays of affection. The owner may benefit practically, through the control of vermin; in the UK, the majority of farmers keep several cats whose primary function is rodent control (Macdonald *et al.*, 1987). However, this role is one of decreasing importance for the majority of cats; indeed many modern owners are distressed by their successes in the hunt. It is as companions that many people keep cats, satisfying their urge to nurture, or, in a family context, giving their children intimate contact with a living being. The relationship is not as strong as that between man and dog, in terms of both the amount of time put into the relationship and the degree of emotional investment. Cat owners value many attributes of their pets, including their independence, cleanliness, reliability in returning home, lack of aggression towards people, and other less easily described qualities such as 'femininity' and 'distance' (H. Karmasin, personal communication).

General Biology

Skeleton and muscles

Many of the ways in which the cat differs from the general mammalian form are related to its carnivorous lifestyle. Unusually among domestic animals, its skeleton and musculature have been scarcely modified from wild *Felis silvestris*. While the skeleton adopts the general mammalian pattern, there is clear evidence of modifications to permit efficient hunting. The most extreme affect the use of the front legs, which are made highly mobile by almost complete reduction of the collar-bone or clavicle, which is replaced by powerful muscles. This allows considerable fluidity of movement for catching prey by the normal method, which uses the claws of the front feet, and for balancing. Further flexibility of movement is achieved by highly mobile joints between the vertebrae. The hindlimbs are more normally jointed, and are specialized for power, used particularly in jumping rather than running. The muscles of these legs tire quickly, and the fastest gait of the domestic cat is considerably less efficient than the corresponding gait in the cheetah *Acinonyx jubatus*, the only felid that specializes in running down its prey.

Studies of locomotion (summarized in Ewer, 1973) have indicated that the fore- and hindlimbs play very different roles, particularly while the cat is walking. Most of the propulsive effort comes from the hindlegs; at the moment when the front legs hit the ground they are sloping forwards, and therefore act momentarily as brakes. In the later part of each stride, the front legs do provide some forward propulsion, but this only cancels out the initial braking effect. The main function of the forelimbs in walking is therefore to take the weight of the relatively heavy front half of the body, while the hindlimbs provide the net power. The 'fluid' character of the cat's walk is achieved by the form of the synchronization between front and rear legs; as each front leg touches the ground, and exerts its maximum braking effect, the hind leg on the same side is exerting its maximum forward effort. In the fastest gait, the gallop, the braking effect is reduced in a different way; the legs are already moving backwards when they meet the ground, and also the spine flexes at this moment, allowing the rear legs to continue their forward progress unchecked. Three variations on the gallop have been detected in cats. In the transverse and rotatory gallops the limbs never strike the ground simultaneously. They are distinguished by the type of synchronization between fore- and hindlimbs, which strike in the same order (e.g. both right–left) in the transverse gallop, and in the opposite order (e.g. hind, right–left; fore, left–right) in the rotatory gallop. The third type, the half-bound, differs in that the hindlimbs are touched down together, but the forelimbs are not. Some animals, but apparently not the cat, use the full bound gallop, in which the hindlimbs touch down together, followed by both forelimbs together. In all three cases there is one flight phase per cycle, in which all four legs are off the ground. This follows the last lifting of a foreleg, and ends when the first (or both) hindleg(s) strike the ground. Any one of the three gallops may be used at velocities between 2 and 6 m/s, the initial choice seeming to depend on individual preference, although the rotatory gallop may be used the most. Increases in speed are achieved by lengthening the stride, and slightly increasing the flight phase. During extended galloping, cats can switch smoothly from one type to another, and also alternate which limb is leading and which trailing, by increasing or decreasing the time taken to lift each limb from the ground (Wetzel *et al.*, 1977). It has been suggested that by switching between the types of gallop, fatigue in individual limbs is minimized.

The skull is notable mainly for its large eye-sockets, characteristic of a visual predator, and modifications to the teeth. There are only 30 of these, fewer than in many other carnivores (the minimum number in the Felidae is 26), and they are almost all adapted to meat eating (Fig. 1.3), with the exception of the incisors at the front of the mouth, which are very small and are used mainly in grooming. The long, laterally compressed canines are used in holding food, and specifically for dislocating the vertebrae of prey. These teeth are equipped with abundant mechanoreceptors, which may be used to sense the precise place that the killing bite should be delivered.

Cats are unable to chew in the way that herbivores do, since their last

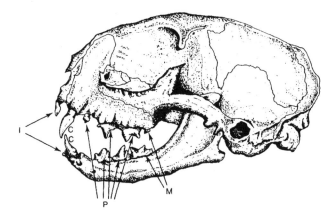

Fig. 1.3. Side (lateral) view of the skull of the cat, showing the arrangement of the teeth. I = incisors (three upper and three lower on each side); C = canines (one upper, one lower); P = premolars (three upper, two lower); M = molars (one upper, one lower). (M.T.)

upper premolars and lower molars, the carnassial teeth, act like shears to cut meat into swallowable pieces. The masseter muscle provides the power for this slicing action, performed when the mouth is almost closed. Considerable power is also required for the capture of prey, when the canines are being used, but under these circumstances the mouth is fully open, the masseter is not particularly efficient, and a different group of muscles (the anterior temporales and the zygomatico-mandibularis) provide most of the power (Fig. 1.4).

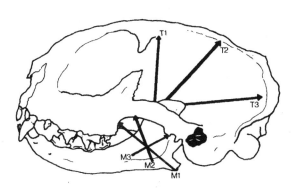

Fig. 1.4. Diagrammatic view of the skull, showing the main jaw muscles as arrows. T1, T2 and T3 are parts of the temporalis muscles; M1, M2 and M3 indicate areas of the masseter muscle. (M.T.)

Skin and coat

The skin is loose, increasing the possibility that wounds incurred in fights with other cats or large prey will be superficial. The wild-type coat consists of long, coarse primary ('guard') hairs, growing singly from follicles, and a variety of shorter secondary hairs growing in groups. Selective breeding has resulted in many modifications, including long guard hairs (Angora), long primary and secondary hairs (Persian), and absence of the guard hairs (Cornish Rex). Muscles on the roots of the guard hairs allow the coat to be erected, either to increase the apparent size of the body in social conflicts, or simply as additional insulation in cold weather. The heavy coat creates problems for thermoregulation in hot weather; eccrine sweat glands are only to be found on the soles of the feet, and cats frequently use the evaporation of saliva to lose heat, either by panting, or by smearing saliva on the coat. To conserve energy, panting occurs at the natural resonance frequency of the respiratory system, about 250 cycles/min. Grooming is usually carried out efficiently, cleanliness being one of the qualities that owners appreciate in their cats. Cleaning, whether direct or indirect via brushing with the front paws, is ultimately carried out by abrasive hooked filiform papillae on the centre of the tongue.

The claws are made from the structural protein keratin, and are derived from the skin rather than the skeleton. Often referred to as retractile, they are more properly termed protractile, since their resting state is sheathed. Each is attached to the final toe-bone, and is unsheathed by tendons which pull on that bone to pivot the claw forwards and downwards.

Reproduction

Many details of reproduction in the cat conform to the standard mammalian pattern, the main difference being the trigger for ovulation. In common with other felids and some other mammals such as the rabbit, the cat is an induced ovulator. Stimulation of the vagina during mating results in a surge of luteinizing hormone from the pituitary, whereupon the mature follicles burst, releasing their eggs. It has been claimed that this is an adaptation to infrequent contact between the sexes, allowing synchronization of ovulation and fertilization, but this is an unlikely explanation for the same phenomenon in the rabbit, a highly social species, and the true value of this system may be found in the mating system of the cat, which is currently under intensive study.

Conclusion

The domestic cat still carries all of the morphological characteristics that were evolved by *F. s. lybica* as a hunter of small mammals in North Africa, and has undoubtedly retained many of its behavioural characteristics as well. Its spread to temperate climates has depended on shelter provided, deliberately or accidentally, by man, and this has been particularly important for ensuring the survival of kittens, which are still susceptible to the effects of cold. The behaviour of the domesticated varieties of *F. silvestris* suggests that they are unlikely to have been derived from the European Wild Cat, and the occidental breeds have almost certainly been derived from the North African race. Given that the domestication of the cat has been a gradual process, there is no reason to suppose that it has not occurred several times, each in a different area. The possibility that the oriental breeds may have been derived from the Indian Desert Cat *F. s. ornata* could explain the behavioural differences between eastern and western breeds. The oriental breeds carry a high frequency of certain deleterious genes, such as those that cause deficiencies in vision (see Chapter 2), which suggests that they have passed through a population bottleneck at some stage post-domestication, as has also happened, under natural conditions, to the cheetah (Kitchener, 1991).

Sensory Abilities

A full appreciation of the way that an animal reacts to its environment depends on our understanding of which stimuli the animal can actually perceive. There is a tendency, that has to be resisted, to assume that mammals see, hear, smell and feel in the same way that we do, and therefore live in the same subjective world that we do. There are obvious exceptions that are generally allowed for, such as the echolocating abilities of bats, but it is easily forgotten, because of their very familiarity, that our domestic animals have different sensory abilities to our own. The domestic cat has been a favourite subject for investigations of the workings of all the major senses, with the exception of olfaction, but it is difficult to find accounts which relate the cat's behaviour and ecology to its sensory abilities, and the perceptions of the world that result from those abilities. This chapter attempts a synthesis of the available literature, so that we can begin to grasp what information the cat can most easily gain about its surroundings, and the extent to which that information overlaps with what we would gather ourselves if placed in the same situations.

The Vestibular System

Our own sense of balance acts almost entirely at the subconscious level, and therefore we do not give it the same level of importance as sight, sound, touch or smell. However, one only has to look at the degree of control that the hunting cat brings to the movement of its body, and especially the movement of its head, to realize that balance has a great part to play in the success of the cat as a predator. The functioning of the principal organ of balance, the vestibular system, is now quite well understood

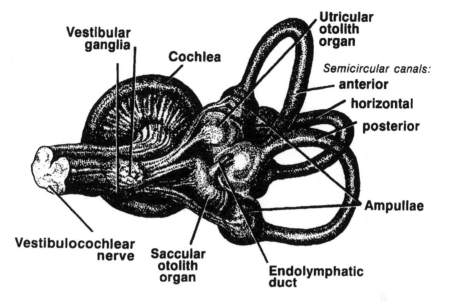

Fig. 2.1. The inner ear, showing the cochlea, which is concerned with hearing, and the semicircular canals and otolith organs which are concerned with balance. Both are connected to a single sensory nerve, the vestibulocochlear, which connects to the medulla in the brain (not shown). (Redrawn from Crouch, 1969.)

(see Wilson and Melville Jones, 1979). A brief description of this organ follows, together with the special features that are known for the cat.

The vestibular system forms part of the inner ear, and consists of fluid-filled tubes of two types. The three semicircular canals detect angular movement of the head in all three dimensions, although the canals are not precisely arranged at right angles to one another, nor are the corresponding pairs of canals on each side of the head all parallel to one another (Fig. 2.1). The canals work on the principle that the fluid tends not to move when the head makes sudden changes of angle, and this relative motion, detected on the walls of the canals, provides sensory information about the change. The other type, consisting of the utricular and saccular otolith organs, primarily detects both linear motion and gravity. A ciliated sensory epithelium is covered by a layer of crystalline deposits of calcite, known as otoconia, which get 'left behind' as the rest of the head accelerates; this deforms the bunches of cilia, which provide the sensory input.

In the cat, the alignment of the three semicircular canals is much closer to being at mutual right angles to one another (orthogonal) than in other mammals, for example man or the guinea-pig. Presumably this makes the integration of information coming from the three canals simpler; if the canals are not at right angles then any motion will result in signals from at least two canals. Furthermore the horizontal canal is precisely aligned with the normal position in which the cat carries its head. The utricular otolith organ is also 'tuned' to measure gravitational deviations from the normal head carriage most accurately. The role of the saccular otolith organ is less clear, but it does not seem to be particularly sensitive to gravity, so it may be specialized as a detector of movement.

The very precise carriage of the head in the cat, also seen in many other carnivores, and notably in birds of prey, is largely a result of integration of information coming from the vestibular organ, and translation of that information into precisely defined movements. While the more complex aspects of this behaviour are under the direct control of the brain, much of the output from the vestibular organ results in reflex movements of the neck and body muscle systems, and of the eyes. These reflexes, and their integration, are described in the next chapter.

Cutaneous Sensory Mechanisms

In common with the other senses, the sensory receptors on the skin of the cat are similar in structure to those found in other non-primate mammals. Their neurophysiology is quite well understood, as are the mechanisms whereby the information they produce is transmitted through the spinal cord and the cerebral cortex (Iggo, 1982). The cat's perceptual world cannot differ very much from our own as far as touch is concerned, but a few special features can be discussed as part of a general description of the types of receptors that are present, and the kind of information that they produce.

There are probably as many as 15 different types of cutaneous receptor, or afferent unit, in the cat. They can be broadly divided into three categories: mechanoreceptors, sensitive to touch and pressure; thermoreceptors, sensitive to temperature, and nociceptors, which in man produce subjective sensations of pain. Each group can be distinguished according to the class of stimulus which will produce maximum discharge, and can also be subdivided by the optimum quality of stimulus. In most cases it is also possible to identify these neurophysiological categories with morphological characteristics, such as structures associated with the nerve endings, and diameters and state of myelination of the afferent fibres. From a behavioural point of view, the most important aspect is the quality of stimulus that produces the response, for it is the sum of these qualities that defines the limits of the perceptual world of the cat.

The mechanoreceptors can be divided into two broad types. One responds mainly to movements of the skin, or hairs with which they are associated, but not to sustained displacements. The other responds to both movements and displacements, for example sustained pressure as is felt on the pads of the feet. The former are known as rapidly adapting (RA) units, the latter as slowly adapting (SA). The resting rate of discharge from the SA units is proportional to the amount of indentation of the skin, or the extent of displacement of a hair or group of hairs. Type I SA units are often grouped in domed 'touch corpuscles', and where they terminate are totally enclosed in specialized structures called Merkel cells (Iggo, 1966). These units are particularly sensitive to stroking of the skin. The other common type of SA unit, SA II, is more widely distributed and terminates in Ruffini endings. This unit is more sensitive to stretching of the skin. A third type of SA unit, called the C-mechanoreceptor, signals lingering mechanical stimulations (those that stay in contact with the skin for 200 ms or more); in the cat, these receptors are more common on the hindquarters than the forequarters, but their function is not yet fully understood.

The RA units respond while displacements of the skin or hair are taking place, but become silent if the displacement is held steady. The hair follicles are innervated by their myelinated afferent fibres, which end in a ring around the sheath of the hair root. There are at least three types of hair follicle unit, each associated with a different type of hair (Table 2.1). The most sensitive are type D, which can be excited by movements of the down hairs alone; the receptive field of each unit can be as much as 2 cm^2, so each unit cannot individually give very precise spatial information. Type G cells are each associated with ten or more guard hairs, and each guard hair can be connected to several different units. The third classification, type T, is not so common in the cat as it is in the rabbit (Table 2.1); each unit is connected to between three and ten of the longest guard hairs, called tylotrich follicles. Pacinian corpuscles, which are encapsulated receptors in the deeper skin layers, are found in both hairy and hairless skin. These are particularly sensitive to vibrations between 200 and 400 Hz, but in common with other RA receptors do not respond to steady-state displacements. This accounts for the cat's ability to detect vibrations through the pads of its feet even while those pads are being distorted by the weight of its body.

Different areas of the skin contain different numbers and proportions of mechanoreceptors. In the cat the nose and the pads of the front feet are particularly well supplied, reflecting the use of the front paws in hunting and manipulating food, although the paws of more manually dextrous carnivores, such as the raccoon, are even more highly innervated. Between the foot and toe pads of the cat are the highest densities of hair receptors anywhere on the body, and the labrous skin is densely packed with several types of receptor. These include two types of RA (velocity-sensitive) receptors, one giving little positional or directional information, and the other indicating both the position and direction of the stimulus (Burgess and Perl, 1973). These receptors are connected together in such a way as to produce tremendous sensitivity to speed and direction of movement of a

Table 2.1. Characteristics of mechanoreceptors abundant in the hairy skin of the cat and the rabbit (saphenous nerve).

Type of unit	Receptive field	Number of units	
		Cat	Rabbit
Rapidly adapting hair follicle afferent units			
Type T	3–10 largest guard hairs	13	41
Type G	> 10 guard hairs	217	46
Type D	guard and down hairs; large field, up to 2 cm^2	89	91
Slowly adapting units			
Type I (epidermal)	touch corpuscle; 1–5 per axon	113	77
Type II (dermal)	spot-like; excited by stretching	39	9

From Iggo, 1966.

stimulus across the pad. The system does not, however, discriminate between textures presented statically, and it is thought that the cat gains information about surfaces by making sequential comparisons as it moves its paw over objects (Gray, 1966). There are no C-mechanoreceptors in the glabrous skin, and all of the SA receptors are type I, although they are not grouped in touch corpuscles, as they are in hairy skin.

In addition to all this innervation of the pads and feet, there are highly sensitive and specialized SA cells in the soft tissue at the bases of the claws. These produce signals about the degree of extension and sideways displacement, which are separate for each of the claws (Gordon and Jukes, 1964). Indeed, it is almost possible to think of the feet as sense organs in their own right, and their degree of sensitivity may explain why many cats appear to dislike having their paws touched.

High concentrations of sensory units are also found around the vibrissae, which are the familiar stiffened sensory hairs found on the head. They are well supplied with both RA and SA mechanoreceptors, so that information on the amplitude, direction and rate of displacement can all be sent to the CNS, and in addition the SA receptors are arranged so that information on static displacement is produced (guard hair receptors show little positional response). Similar arrangements of receptors are found at the bases of the shorter stiffened hairs around the lips. Other clusters of stiffened hairs, the carpal hairs, are also found on the wrists; in addition to SA and RA receptors, these are equipped with vibration-detecting Pacinian corpuscles, which the vibrissae are not (Burgess and Perl, 1973).

Both the vibrissae and the carpal hairs provide sensory information about the position of the cat's head and legs in relation to nearby objects,

Fig. 2.2. Arrangement of the vibrissae on the head, drawn diagramatically to show the position of each tuft. M = mystacials, S = superciliary, G1, G2 = genals. (M.T.)

and may also be sensitive to air currents. This information may be most useful in the dark, or when the cat is manoeuvring in a confined space, and may help to compensate for the cat's long-sightedness when objects are being manipulated close to the snout. The vibrissae are arranged in tufts, of which the largest are the mystacials or 'whiskers' (Fig. 2.2). These can be moved backwards for protection, or forwards towards objects to be investigated. The sensory input from the whiskers to the brain is co-ordinated with visual information in the superior colliculus (Fig. 2.3). The large amount of nervous tissue devoted to processing and integrating information coming from the mystacials indicates their importance as sense organs, perhaps most importantly when the snout is being directed towards prey.

The other tufts are the superciliary, which act like extensions of the eyelashes in triggering the protective eyeblink reflex, and the two genal tufts on the cheeks, which are close to enlarged skin glands, and may act as scent spreaders as well as tactile sense organs (Ewer, 1973). Many carnivores also have a tuft of vibrissae under the chin, the inter-ramal tuft, but this is absent from the cat family. Since cats do not tend to hunt with their noses to the ground, or dig with their noses, they may not require these vibrissae to the same extent as other carnivores. It is also possible that stiffened hairs in this area would interfere with the dispersion of scent from the submandibular glands (see Chapter 5).

The remaining types of cutaneous sensory units respond to temperature (thermoreceptors), and to severe mechanical and thermal stimulation (nociceptors). There are two basic types of thermoreceptor – warm receptors with non-myelinated afferent fibres, and cold receptors with myelinated fibres. Both respond to absolute temperatures, and to changes in

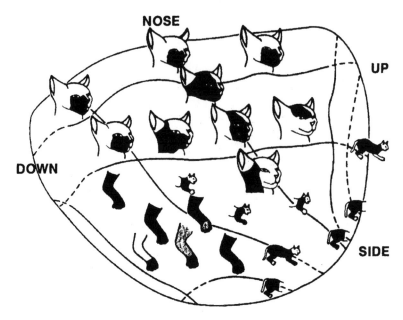

Fig. 2.3. A diagram of the surface of the left superior colliculus, part of the mid-brain that processes and integrates sensory information. For each part of the surface, diagrams indicate from which area of the skin (shown shaded) tactile information is processed. The relative sizes of the diagrams indicate the amount of nervous tissues devoted to each area, and illustrate the importance of touch sensations coming from the face and the forepaw, and the relative unimportance of the body and the hindlimbs. The contour lines indicate from which direction each part of the superior colliculus processes visual information. The horizontal plane, as seen by the eye, is indicated by the line running from NOSE (i.e. towards the nose) to SIDE (i.e. to the side), and the vertical plane by UP to DOWN. It can be seen that the visual and tactile information is approximately superimposed. For example, the part processing tactile information from the top of the head also processes visual information from the upper central part of the visual field; touch sensations from the forepaws are processed by the same part in which their visual image would normally be represented (down and out to the side of the field of view). (Redrawn from Stein *et al.*, 1976.)

temperature. The warm units are maximally stimulated by steady temperatures of 40–42°C, and also increase their rate of firing as the skin temperature increases. The cold units increase their firing rate as the temperature drops, and respond most powerfully at steady temperatures of 25–30°C.

Discharge of nociceptors is associated with sensations of pain in human subjects, but their presence in the skin of the cat does not necessarily mean that cats feel pain as we do. These receptors fulfil a protective function, necessary because both mechanoreceptors and thermoreceptors reach their maximum discharge rate (and can therefore give no further information) to stimuli that are intense, but not severe enough to be potentially damaging. Nociceptors consist of unencapsulated nerve endings of two types, a mechanical type that responds to squeezing and crushing of the skin, and a mechanothermal type that also responds to extremes of heat and cold. Nociceptor responses are probably poorly localized, due to convergence of the neurone, but spatial information may be provided by the other cutaneous receptors that are inevitably discharged at the same moment.

Hearing

The detection of vibrations in the ground may be primarily detected by the Pacinian corpuscles in the feet, and at the bases of the carpal hairs, but airborne vibrations are detected as sound by the hearing system. This broadly conforms to the general mammalian pattern, both in structure and function. The ears are commonly thought of as consisting of three units. The outer ear, comprising the pinna and the canal, channels the sound, still consisting of airborne vibrations, to the middle ear (Fig. 2.4). This consists of the ear drum or tympanum, and the ossicles, small bones that transfer variations in air pressure into variations in fluid pressure in the inner ear. These vibrations are picked up by a variety of nerve cells, some tuned to particular frequencies, and some to specific changes in frequency. The former permit the discrimination of two frequencies presented simultaneously (to take an example from music, the constituent notes of a chord). The latter can be tuned to species-specific sounds; for example, click-specific detectors are used by bats in echolocation. In man, both the outer and middle ears are structured so as to boost those frequencies that are particularly important in speech; the cat's pinna functions as a directional amplifier, boosting frequencies in the range 2–6 kHz (Martin and Webster, 1989), and it may be that this, and some other aspects of the cat's hearing, are adaptations to the detection of species-specific calls, although this does not seem to have been investigated in detail. The bulk of the evidence suggests that many aspects of hearing in the cat have been shaped by the need to detect prey.

The detection of sounds

The audiogram of the cat, its ability to detect pure tones over a range of frequencies, is about 10.5 octaves, among the broadest recorded from any mammal; only the horse and the porpoise have slightly larger ranges (Fay, 1988). For comparison, that of man is about 9.3 octaves. The structure of the cat's head, and the distance between its outer ears, have been compared with that of other mammals, and these comparisons predict that the

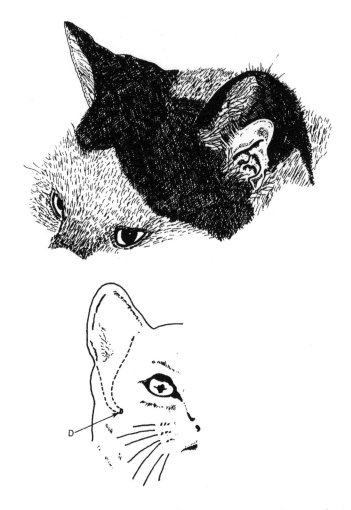

Fig. 2.4. The external ear (pinna) showing the complex corrugations on the inner surface that assist in the location of sources of sounds. The position of the middle and inner ears are shown below. D = ear drum. (M.T.)

hearing range has been extended at both the high-frequency and low-freqency ends. By what precise means is still uncertain, since most mammals appear to have 'traded off' high-frequency ability for low, or vice versa (Heffner and Heffner, 1985).

At low frequencies (above 50 Hz), the cat's thresholds are broadly similar to those of man, a remarkable ability considering the much smaller head of the cat. The biological function of this ability has not yet been elucidated. In the middle frequencies (1–20 kHz), the cat is one of the most sensitive mammals; some studies have indicated thresholds down to

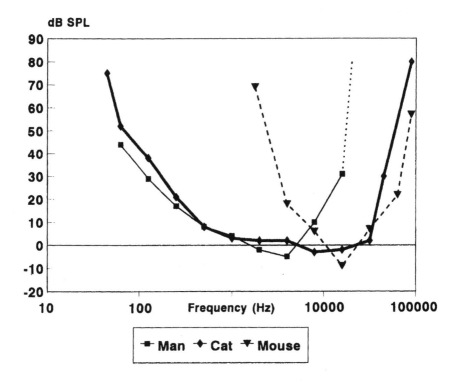

Fig. 2.5. Auditory thresholds for the detection of pure tones, in man, cat and mouse. The thresholds are expressed as decibels of sound pressure level (SPL); the lower the figure, the more sensitive the hearing at that frequency (high notes have high frequencies, and the region above the human range is normally regarded as ultrasound). Note that the hearing ranges of man and mouse hardly overlap at all, but that of the cat overlaps substantially with both. (Data from Fay, 1988.)

between −20 and −25 dB sound pressure level (SPL), compared to −5 dB for a young adult human, although others, such as that illustrated in Fig. 2.5, have not. Reproducible differences can be detected in this region between the abilities of individual cats, which may explain some of the discrepancies. Some studies have detected a slight diminution in sensitivity, of unknown significance, at about 4 kHz. While man's thresholds decrease above 4 kHz, the lowest thresholds are at around 10 kHz for the cat, which can still detect sounds at 85 kHz if they are reasonably loud (the human equivalent would be louder than normal conversation but quieter than shouting). Because there is no abrupt cut-off in sensitivity at high frequencies, it is difficult to quote an exact figure for the upper limit, but it is generally accepted that the useful limit for the cat is about 60 kHz. This ability is particularly remarkable because the efficiency of the cat's middle ear drops at frequencies above 10 kHz, due to mass limitations in the ossicles, and to 'break up' of the ear drum into smaller vibrating portions. Unlike many mammals, adult cats are not known to make ultrasonic calls, and so this high-frequency ability is presumably related to the detection of such calls emitted by small rodents, and is therefore an adaptation to hunting by sound cues.

In the ability to discriminate between sounds of the same frequency but different intensities, the cat is similar to other mammals, and inferior to man. Man is also better at discriminating pairs of sounds of the same intensity but different frequency, provided the frequency is below 5 kHz. Above this figure, cats outperform man, although significant differences between the abilities of individual cats are evident. However, it should be noted that some of these interspecific differences may be due to the procedures used. The comparisons just made rely on the results of conditioning experiments (cats) and verbal reports (human subjects). Neurophysiological measures from cochlear nerve fibres indicate that cats have the potential to be as discriminating as man at low frequencies, and to be considerably better at frequencies over 2 kHz. Whether the behavioural measures underestimate the cat's true abilities, or whether some discrimination is lost during the processing of the information in the CNS, has not been resolved. Cats also appear to be less sensitive than man in the detection of sounds of very short duration (Costalupes, 1983); man's abilities in this area have been linked to the complexity of our vocalizations.

Localizations of sound in space

All of the types of discrimination considered so far could be done as well with one ear as with two. Cats that are deaf in one ear do learn to locate sounds, often by making exaggerated scanning head movements to produce alterations in the intensity and quality of the sound at the single working ear. However, such cats never approach the abilities of those with complete hearing, showing that comparisons of the sound arriving at the two ears are crucial to the pinpointing of the source of a sound, a skill that is critical to successful hunting in cover. The problem facing the cat can be broken down into the position of the sound source in the horizontal plane, the plane in which its ears are normally separated, and in the vertical plane, in which they are not.

Location of sound in the horizontal plane is achieved by comparison of the different signals arriving at the two ears, but the type of comparison that can be made is an inevitable function of the physics of the sound. For signals below about 4 kHz, for which the wavelength is less than the distance between the pinnae, time comparisons are made; the relatively small head also makes comparisons below 500 Hz difficult. At frequencies above 6 kHz, time differences become ambiguous, and the cat seems to rely on differences in intensity, produced by the masking effect of the head

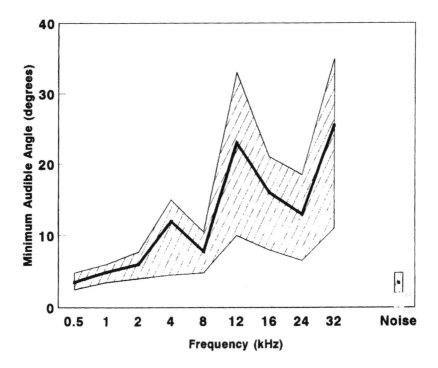

Fig. 2.6. The minimum angles in the horizontal plane at which five cats could distinguish between two sources of sound. The best discrimination, in all individuals, was between noise (mixed frequencies), shown at the right. Low-frequency sounds were also well separated, but there was a general trend towards poorer resolution (solid line), and greater individual variation (shaded area) with higher frequencies. (Redrawn from Martin and Webster, 1987.)

itself, and the directional properties of the pinnae, such that more high-frequency sound reaches the nearest ear. There is a range of frequencies over which neither method is particularly accurate, resulting in a decline in angle-separating ability between 4 and 12 kHz, before the discrimination increases again between about 14 and 24 kHz (Fig. 2.6). Of course, a hunting cat is unlikely to encounter pure tones, and the minimum angle that can be discriminated for noise signals that contain a broad range of frequencies is lower than for any pure tone.

There is still some uncertainty about the role of the pinna in the location of sound. In the human ear, the corrugations add reflections to the spectrum of incoming sounds. These reflections give information on the elevation of the source of the sound, and whether it comes from behind or in front of the head, for which comparisons of the signals from the two ears would be ineffective. However, the advantages that the cat gains from the considerable mobility of its pinnae, particularly in the horizontal plane, are far from clear. Heffner and Heffner (1988) have suggested that the pinnae might be repositioned, either between bursts of an intermittent sound, or during a drawn-out sound. They were able to show that, compared to man, cats are less able to localize sounds that are straight in front of them, but are as good as man when the sounds come from the side. The mobile pinna may therefore help to spread the localizing ability over a wide range of angles. However, there is a cost involved, since for a mobile pinna the CNS will have to make direction-dependent adjustments for the position of the pinnae in analysing the differences between the signals reaching each ear, much as the visual system corrects directional information for the position of the eyes in the sockets. Most cats find localization of sound in the vertical plane almost impossible if the sound consists of a single frequency, but are very accurate if the sound is of mixed frequencies (Martin and Webster, 1987). This suggests that the cat, like man, uses spectral trans-formation cues to determine elevation. At frequencies above 12 kHz, the overall shape of the pinnae, and the corrugations within it, produce complex differences in the signals reaching each ear – for some combin-ations of elevation and frequency the signal is actually louder in the more distant ear (Martin and Webster, 1989). By comparing the intensities of different frequencies, the cat has sufficient information to localize the height of the source of the sound accurately. This ability must presumably be learned, as the shape of the pinna varies from one individual to another, and changes as the cat grows. The information must also be corrected for rotation of each pinna, as discussed for horizontal location; it appears that so far more costs than benefits have been found for having a mobile pinna, so further aspects of sound location by the cat undoubtedly await discovery.

Vision

The eye

The physiology of vision is probably better understood in the cat than in any other non-primate mammal, and yet the behavioural significance of this physiology, and of the several ways in which vision in the cat diverges from the general mammalian pattern, is still imperfectly understood. Table 2.2 illustrates some of the differences between the structure of our own eye and that of the cat. Broadly speaking, the cat is much better adapted to vision at very low light intensities than we are. The maximum density of rods, the most sensitive visual receptors, is almost three times our own, while there are far fewer cones, the less sensitive visual receptors that we use in daylight. The centre of our own visual field corresponds on the

Table 2.2. A comparison between ocular and retinal measurements for the cat and man.

Parameter	Cat	Man
Ocular		
Eye diameter (mm)	22.3	25
Total power of ocular optics (D)	78	60
Pupil		
Maximum diameter (mm)	14.2	8
Maximum area (mm^2)	160	51
Minimum diameter (mm)	<1 (width)	2
Interpupillary distance (mm)	36	62
Retina		
Maximum cone density (mm^{-2})	26×10^3	146×10^3
Minimum intercone distance (μm)	6.2	2.5
Maximum rod density (mm^{-2})	460×10^3	160×10^3
Maximum ganglion cell density (mm^{-2})	3×10^3	147×10^3
Optic nerve and chiasma		
Total number of fibres ($\times 10^3$)	85	1100

From Berkley, 1976.

retina to the fovea, an area in which rods are absent and cones are present at high density. Cats have no fovea; in their corresponding area centralis the density of cones is six times lower, and many rods are also present. The innervation of the eye is also differently organized; man's optic nerve contains over ten times more fibres than that of the cat, and there is an even greater discrepancy in the maximum density of ganglia in the retina itself. Overall, this results in more rods and cones being connected to each nerve, an adaptation towards greater sensitivity and therefore better night vision; this adaptation has its costs, as we shall see. The efficiency of vision is still further enhanced by the tapetum, a layer of reflective cells immediately behind the retina, which reflects back any light that has not been absorbed by the visual pigments in the rods and cones at the first pass, giving a second chance for absorption before the light passes back out of the pupil again. It is this process that gives cats their 'eye-shine' when a strong light is shone at them when it is dark and the pupil is wide open. Tapeta are found in a variety of vertebrates; the type found in the cat, a tapetum cellulosum, is also found in some other terrestrial carnivores, seals, and some lower primates. The reflecting material consists of rodlets, 230 nm in diameter, arranged in regular arrays within cells in the choroid. The reflective properties seem to be due to the presence of riboflavin. In the cat, the tapetum increases the efficiency of the eye by about 40%.

Cats have large eyes in proportion to their body size, and the path of light from the pupil to the retina is shorter than in man, increasing efficiency further. Their pupils can be opened wider than can ours (Table 2.2), but in order to protect the sensitive retina, and to allow vision under bright conditions, they also have to be closed to a smaller area than ours need to be. A simple circular arrangement of muscles around the iris appears to be inadequate to generate this extreme range of pupil sizes, and instead of contracting from a large circle to a small one the cat's pupil contracts to a fine slit, less than 1 mm wide. When the pupil is fully open, retinal illumination is of the same order as that of nocturnal creatures, such as the badger and bats. Diurnal primates, such as man and the chimpanzee, have about five times less retinal illumination (Hughes, 1977).

There is considerable disagreement in the literature on the abilities of cats to transfer the focus of their eyes from near to distant objects, and vice versa. However, all the measurements agree in awarding man superior abilities. Some of the discrepancies within the data for cats may relate to differences between cats reared indoors, and those allowed access to a variety of environments from an early age. Feral cats tend to be slightly long-sighted, whereas cats raised indoors, where the furthest objects on which they can focus are only a few metres away, become short-sighted (Belkin *et al.*, 1977). When the cat is completely relaxed, for example during sleep, the eye is long-sighted, and has to be brought into focus on awakening. The normal degree of accommodation in the awake cat should produce a reasonably clear image of any object more than 2 m away, and the further degree of accommodation that has been observed, up to about 4 diopters, should produce clear vision at about 25 cm. The behavioural evidence suggests that cats cannot focus on objects closer than this (Elul and Marchiafava, 1964). Some stimuli elicit much more accommodation than others; a clump of feathers is far more effective than a mouse or a kitten at the same distance. Moving stimuli are also more effective than stationary ones, suggesting that cats may only focus their eyes accurately if detailed information is required.

At least in adult cats that have been reared indoors, the lens is stiff. Therefore, distortion of the lens, the main focusing method in the human eye, is unlikely to be effective for the cat. When cats are looking at objects close to their faces a bulge can be seen on the iris, suggesting that most of the accommodation to distance is achieved by the whole lens being moved backwards and forwards.

The eyes are placed on the head so as to point at about 8° off the centre line; the eyes of most mammals diverge much more than this, although felids tend to have small divergences (the lion's is about 5°). This restricts vision to the area in front of the head (Fig. 2.7), an adaptation to a carnivorous lifestyle, since animals that are preyed upon need all-round vision for protection. The total visual field extends to about 200° in total, with

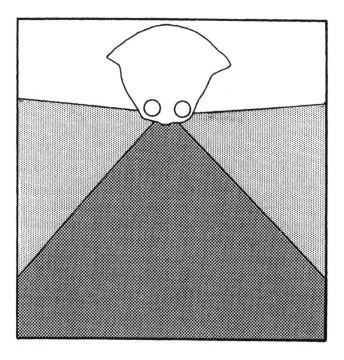

Fig. 2.7. The field of view, including the binocular area (heavily shaded). (M.T.)

binocular vision in the central 90–100°, figures not dissimilar to those for human vision. Within the region of binocular overlap, we can assume that the images are fused, so that cats have single vision. The eyes can be seen to make conjugate movements as if each is fixated on the same object, and identical changes in focus occur in both eyes even if only one eye can actually see the object that is being focused on. As objects are brought closer to the face, vergence movements can be observed, that is, the eyes are brought into a slight squint. Feral cats are reported to display more accurate vergence movements than cats reared indoors, so developmental factors may have a part to play in allowing binocular vision at close range (Hughes, 1972). The limit of vergence for many cats seems to be between 10 and 20 cm, over which range focus is probably also poor. This suggests that vision plays little part in directing the close-range manipulation of prey, for which tactile stimuli are probably much more important, accounting for the cat's sensitive vibrissae.

Despite their relatively large eyes, cats are able to make very rapid eye movements in response to moving objects. If an object moves very quickly, or suddenly appears off the centre of the cat's line of sight, characteristic movements of the eye, known as saccades, are made. These consist of a lag phase of about a quarter of a second, during which the sensory information is translated into instructions for the ciliary muscle, a rapid phase of acceleration, and then a longer deceleration phase. While the saccade is going on, the position of the object is continuously monitored, and if it moves again a second saccade can immediately follow the first to bring the eye back on to the target. That these second saccades are usually successful is a tribute to the ability of the nervous system in making relative calculations of the position of the object on the retina, and its actual position in space. Horizontal and vertical saccades are brought about by different pairs of extraocular muscles, and are controlled by different areas of the brain stem. The vertical movements are the faster (velocities of up to 250°/s have been recorded); the horizontal movements reach a maximum speed of about 100–150°/s over a wide range of total angular displacements. If the object that is followed moves diagonally, the corresponding saccade proceeds at the pace of the slower component, i.e. the horizontal.

If the change in angle is small, about 8° or less, a saccade may occur, but cats have an alternative slow eye movement, lasting over twice as long as a saccade, that has not been observed in primates. The biological significance of these alternatives is unknown. The slow eye movements are also distinct from the smooth eye movements used to track slowly moving objects. Cats are much poorer at this than are primates, and tend to lag behind the object, catching up with small saccades, at speeds as slow as 2°/s. If the whole background moves, as it inevitably appears to do when the head is rotated, the maximum speed of slow eye movements increases to about 8°/s but only in the horizontal plane. The maximum is slower upwards, and much slower downwards; abrupt downward movements of the visual field occur during normal walking, so the latter may be necessary to prevent exaggerated eye movements being triggered by locomotion (Evinger and Fuchs, 1978).

Visual abilities

The absolute threshold for detection of light by the cat is about 3×10^{-7} cd/m^2, about three to eight times better than man's. This difference is almost entirely due to the greater efficiency of the cat's eye, as described above. In both cat and man, about 10 quanta of light, spread over the whole retina, are enough to cause a response; in other words, both systems are reaching the limits of sensitivity as determined by their biophysics. Cat and man are also similar in their sensitivities to different wavelengths at low light levels (i.e. when rods, and not cones, are active); the only small disparity is a small peak at 560 nm in the cat's spectrum, which corresponds to the frequency most efficiently reflected by the tapetum. However, our own vision is superior in several respects to that of the cat. For example, in moderate lighting cats are less able to discriminate which is the brighter of two lights, although at low light levels there is less disparity (Berkley, 1976).

At the onset of darkness, the human eye takes 30–40 minutes to become fully sensitive, due to the necessity for biochemical processes in the visual pigments. In the cat, such total adaptation can take up to one hour if the pupil has been artificially dilated, but it has been suggested that cats may protect their dark adaptation by constricting their pupils when the light level increases suddenly. Of course, the timing of dark adaptation is unlikely to reduce the efficiency of vision at any time of day under natural light, because at dusk and dawn light levels change relatively slowly; it is our use of artificial light that has brought this inadequacy to our, and presumably our cats', attention.

Under conditions of moderate light intensity ($2-75 \text{ cd}/\text{m}^2$), cats pick out far less detail than we do. Their threshold for distinguishing striped black-and-white patterns from uniform grey is about five stripes per degree of arc, while we can distinguish patterns that are six times finer. This can be attributed to at least three separate factors: the scattering of light by the tapetum, the smaller number of cones on the cat's retina, and the greater numbers of rods connected to each retinal ganglion (Pasternak and Merigan, 1980). If the striped patterns are dark grey on light grey rather than black on white, the best patterns for detecting contrast can be measured; in the cat these are about one stripe per degree of arc. Contrast-detecting cells can be found in the striate cortex (area 17 – see Fig. 2.8), each tuned to a particular frequency, and most abundant around the behaviourally measured optimum frequency (Woodhouse and Barlow, 1982).

There are various other standard psychophysical measures that can give an idea of the amount of detail that the cat's visual system can resolve. The Critical Fusion Frequency (CFF) measures the point at which a flashing light is seen as constant illumination, a critical factor in our own perception of films and television. At low levels of illumination ($0.03 \text{ cd}/\text{m}^2$) cats can just distinguish 60 cycles/s as flickering, better discrimination than our own, even when the lighting level is corrected for the cat's more efficient eye. This means that many cats are likely to see fluorescent tubes and televisions as giving off a flickering light. However, the cat's perception of slower flickering can be poorer than our own, particularly if, instead of the light being extinguished between bursts, it is merely made dimmer (Berkley, 1976).

Such temporal responses are presumably linked to the ability to detect movements and shapes. Considerable progress has been made in understanding the neurophysiology of the transformation of the simple on/off responses of the rods into the detection of particular shapes and patterns of movement. As the information moves along the visual pathway to the visual cortex, the neurones respond to increasingly sophisticated combin-

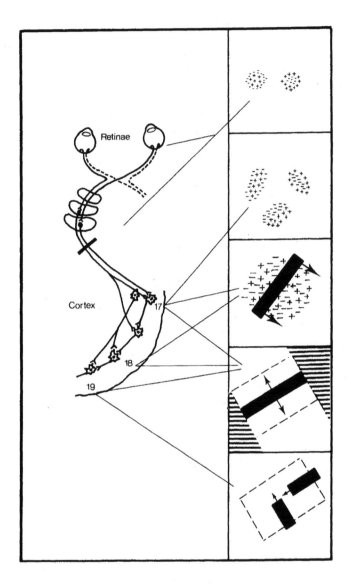

Fig. 2.8. The pathways of visual processing. The ganglia in the eyes (top), and in the lateral geniculate nucleus, shown below the eyes, respond to local brightening and dimming (box, top right). Those in area 17 of the visual cortex respond to moving slits, bars, or edges with specific orientations (upper centre box). Areas 17–19 also contain ganglia that respond to more complex trigger features, such as bars or edges moving in specific directions (centre box), moving edges of a specific width (lower centre box), and blocks of a specific size moving in specific directions (lowest box). (M.T.)

ations of light and shade (Fig. 2.8). In the visual cortex, the majority of neurones are directionally selective, and many are also sensitive to the orientation of the stimulus, and even to particular velocities of movement. Behavioural measures of the cat's abilities in this area have lagged behind, but several studies have indicated a sophisticated level of visual information processing. It is known, for example, that cats can discriminate small differences in the shapes of geometric figures such as triangles. They are also able to make relative size judgements; when trained to discriminate the larger of a pair of figures, the larger was still picked out even when the absolute size of both was altered. White-on-black shapes are seen as similar to the black-on-white version of the same shape (Berkley, 1976).

The detection of movement is an obvious necessity for hunting, but cats seem rather poor at detecting slow movements. The slowest angular speeds that they can be trained to detect are between 1° and 3° of arc per second, whereas we can detect speeds about ten times slower. The direction of movement, up, down, left, right or diagonal, has little effect on the threshold (Pasternak and Merigan, 1980). The finest distinctions of velocity can be made at speeds of 25–60°/s, which presumably correspond to the speeds at which prey are likely to move, since accurate estimation of angle and speed must be crucial to a successful pounce.

Recently, an investigation has been made of the ability of cats to join partly hidden outlines together, thought to be a component of the mechanism that allows the separation of 'figure' from 'ground'. The method relied on the illusion shown in Fig. 2.9; in the upper set of figures, the eye constructs an illusory black square that can be made to appear to move down the background. If the white sectors are aligned in other ways, as in the lower set of figures, no such squares result. Bravo *et al.* (1988) have demonstrated with training experiments that cats can discriminate the figures that produce illusory squares from those that do not. Another way that we separate objects from their background is by discontinuities in textures. Cats can discriminate between the two images shown in Fig. 2.10, showing that they too can use texture to segregate images (Wilkinson, 1986). Such experiments suggest that the retinal image is segmented into biologically relevant features at a relatively low level of neural machinery, and that perception of such illusions does not involve cognitive processes. While no simple pattern-recognizer has been identified in the cat, akin to the frog's 'fly-detector' that directly elicits feeding movements, there is now no doubt that the cat's detection of movement and pattern, and their interactions, are highly sophisticated.

Colour vision

Although it is now established that cats do see in colour under daylight conditions, the extent to which colour is meaningful to them is still under

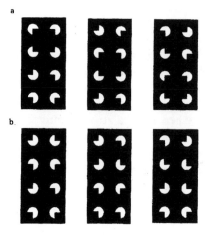

Fig. 2.9. (a) Viewed one after the other, the upper set of pictures appears to show a square that descends from the top set of discs to the bottom, while all the other discs appear to spin at random. (b) In the lower set, all eight discs appear to spin. Cats can discriminate between the visual illusion shown in (a) and the random pattern in (b), indicating that they do see subjective contours. (Redrawn from Bravo *et al.*, 1988.)

debate. There is clear neurophysiological evidence for at least two types of cone, one green sensitive peaking at wavelength 560 nm, and one blue sensitive, at 455 nm. A third type, red sensitive, has been proposed, but its presence has not been confirmed. It is reasonably certain that cats are dichromatic, seeing two pure colours and their combinations, rather than the three pure colours that most people can see. The lack of red-sensitive cones means not only that cats do not see the colour red, but also that red objects appear much darker to them than they do to us, compared to green or blue objects.

Under dim light, the cones are inoperative, and since all the rods are maximally sensitive to light of wavelength 497 nm, cats will see in black and white, as we do. At background illuminations of between 3 and 30 cd/m^2, the cones take over, and two peaks of sensitivity can be detected, one produced by the green- and the other by the blue-sensitive cones. The greater sensitivity is to the blue, and this cone also dominates both temporal and spatial resolution when sufficient light is available (Loop *et al.*, 1987).

Some idea of the unimportance of colour for cats can be gained from the number of wavelength-comparing ganglion cells that can be detected, about 16 times fewer than in primates. This has led to speculation that cats not only perceive fewer colours, but that these colours are much less saturated than those we perceive.

It has proved very difficult indeed to train cats to respond to different colours of the same subjective brightness. Some investigators never succeeded, others demonstrated weak effects after thousands of trials, or many months of continuous training. It is these results that have led to the view that cats are behaviourally colour blind, whatever the physiological evidence. The most likely explanation is, however, that colours provide

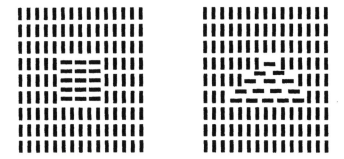

Fig. 2.10. Cats can discriminate between subjective shapes that are formed by changes in texture, such as the examples shown (Wilkinson, 1986).

poor training cues for cats, because of perceptual dominance. This means that cats are much more likely to associate other visual cues, such as pattern or brightness, with the rewards they obtain during training, and therefore take a very long time to realize that the appropriate cue is actually colour. By pairing colour with a pattern, Meyer and Anderson (1965) were able to achieve much more rapid training than when colour was used alone. Red and green colours were overlaid with diagonally striped patterns of opposite orientation; once the cats had learned to distinguish the two, the training continued with the same colours, but with a gradual rotation of the patterns in opposite directions, until they were both vertical and therefore indistinguishable. The cats had by then transferred their training from the pattern to the colour. Other experiments showed that brightness, as well as pattern, is dominant over colour.

Stereoscopic vision

The cat's brain contains binocularly activated nerve cells that respond to retinal disparities, but some authors have maintained that cats do not have true stereoscopic vision. However, cats can distinguish between a large and a small object of the same shape, even when the large object is so far away that its image on the retina is smaller than that of the small object (Gunter, 1951). Although it is theoretically possible that distance to the image could be estimated by the degree of focusing required, we have already seen that the cat's accommodation is weak, and in any case binocular cues are likely to be much more reliable. One-eyed cats can make comparisons of depth, but while doing so they make exaggerated head movements, which would provide a temporal version of binocular vision. They are also much less accurate at judging distance than normal cats. The best evidence is that cats, like us, use retinal disparity to judge depth; that is, they rely on the double images that are formed of objects both in front of and behind the plane of fixation (Fox and Blake, 1971). The value of this ability for hunting is, of course, considerable.

The visual system

The cat has the most extensively studied visual system of any vertebrate animal, but it is not yet possible to integrate all of the abilities described above into a whole that makes biological sense. At this point, knowledge of the neurophysiology is still in advance of the behaviour; for example, pathways for the parallel processing of information about spatial frequency, motion, colour and binocular disparity have been detected, but the resultant effects on behaviour are still poorly understood. However, we now understand a great deal about the visual world in which the cat lives, which is helpful for the interpretation of responses to biologically meaningful stimuli.

It should be remembered that the comparisons that have been made with the abilities of primates, particularly man, often rely on different methodologies for each species, and may therefore be unreliable. Difficulties in training cats to respond to particular stimuli may lead to false conclusions, as for example happened for colour vision. There has also been a tendency to examine aspects of vision in the cat that are familiar from our own visual experience; some of the cat's visual abilities may still be undetected, simply because they have no human counterpart.

Abnormal vision in Siamese cats

Vision in Siamese cats differs from that in non-orientals in several ways, one of which is a reduced ability to detect flicker. In the retinas of all cats two types of ganglion cells can be detected – X-type, which respond to patterns of luminance, and Y-type, which respond to movements, and also inhibit the output of the X-cells while the eyes are moving. Siamese cats have only about 14% of Y-cells, compared to 35–45% in crossbred non-orientals (Loop and Frey, 1981). What these differences mean in terms of the subjective impressions of vision in Siamese cats is difficult to say precisely, but they may be less sensitive to flickering lights, while their vision may be temporarily impaired, possibly by blurring, while they are moving their attention from one object to another.

More serious may be their lack of stereoscopic vision. Most of the nerve fibres from each eye cross over in the optic chiasma and innervate the contralateral side of the brain; in normal cats about 35% of the fibres stay on the same side, enabling comparisons to be made of the disparities between the images received by each eye. Very few binocularly driven cells can be detected in most Siamese cats, and no behavioural evidence can be found to indicate that they have stereoscopic vision, although the acuity of each eye is apparently normal. Probably as a result, Siamese cats have a tendency to develop convergent squints.

The Chemical Senses

There are now acknowledged to be four distinct chemical senses in most mammals: taste, olfaction, the vomeronasal and the trigeminal. Two of these, taste and vomeronasal, have been specifically studied in cats. Since the sense of taste is intimately bound up with food selection and the detection of nutrients, it will be described in the chapter on feeding behaviour. Published accounts of the olfactory and trigeminal senses have tended to focus on other mammals, and so the descriptions that follow necessarily assume that the cat is not very different from other mammals, particularly the domestic dog.

Olfaction

The sense of smell is very important to cats, as it is to the great majority of non-primate mammals. Some idea of this can be gained by the behavioural changes that occur when cats lose their sense of smell as a result of virus infection, particularly feline viral rhinotracheitis. Such cats frequently lose their appetite, change their toileting habits, and do not indulge in courtship. The damage may well extend to the trigeminal and vomeronasal systems as well, so these changes do not indicate that they are triggered by exclusively olfactory stimuli. More direct evidence comes from the size of the olfactory epithelium, which can be 20 cm^2, compared to 2–4 cm^2 in man. This epithelium is supported on some of the scroll bones, the ethmoturbinals, while others, including the maxilloturbinals, serve to clean, warm and moisten the inspired air before it reaches the delicate olfactory receptors (Fig. 2.11). The maxilloturbinals, while moderately well developed in the cat, are even more extensive in more active carnivores such as the dog, where more rapid respiration during exercise will bring greater volumes of air into contact with the olfactory apparatus.

The detailed structure of the olfactory epithelium itself has been revealed by electron microscopy, and is considerably simpler than, for example, that of the retina (Fig. 2.12). The surface is protected from direct contact with the air by a layer of mucus secreted by the Bowman's glands, and through which the molecules of odour must pass before they can reach the receptors themselves. These receptors are carried on the dendrites of the first-order olfactory neurones, the axons of which make contact with the second-order neurones in the olfactory bulb. The detailed structure of the vesicle, the part of each receptor cell that projects into the mucus, gives

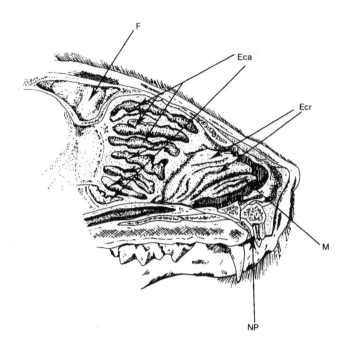

Fig. 2.11. Diagram of the bony structures that support the olfactory apparatus in the cat. The olfactory epithelium is attached to the caudal ethmoturbinals (Eca). Other structures shown are the frontal sinus (F), the cranial ethmoturbinals (Ecr) and the maxilloturbinals (M). The opening of the nasopalatine canal, which leads to the vomeronasal organ, is also shown (NP). (M.T.)

some further indications that the cat's sense of smell is highly sensitive. Most of the individual receptors are carried on ciliary processes that project sideways into the mucus, up to 150 per vesicle, and up to 80 μm in length, much greater than those of most other mammals (Andres, 1969). Between the receptor cells are the supporting cells, which send unusually long, abundant microvilli into the mucus, presumably as mechanical support for the cilia. They appear to be metabolically active, since their cytoplasm contains large concentrations of mitochondria and endoplasmic reticular membranes. Immature forms of these cells can be interspersed with the mature forms, and both support cells and receptor cells are replaced regularly. In the cat, a second type of support cell has also been described, in which the microvilli are straight and stiff.

The olfactory abilities of cats have not been extensively studied, but it is likely that they are similar to those of the domestic dog, which is much more readily trained to carry out the appropriate psychophysical discriminations. Dogs are capable of detecting some compounds at thresholds about a thousand times lower than man's (Davis, 1973), and can discriminate

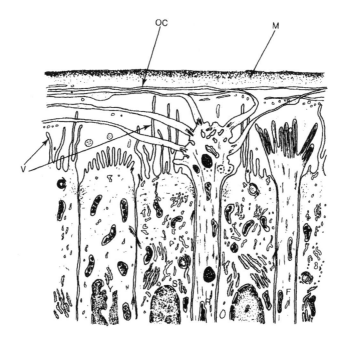

Fig. 2.12. Representation of the cells bordering the olfactory membrane, redrawn from electron micrographs (Andres, 1969). The olfactory receptors are carried on long cilia (OC) that originate from the receptor cells (R); to reach them the odorous molecules must dissolve in, and travel through, the mucus (M) in which they are covered. Mature (S) and immature (Y) support cells occur between the olfactory cells. The 'fifth' cell type (F), apparently a specialized support cell, is not found in the dog. (M.T.)

between identical twins on the basis of odour alone (Kalmus, 1955). Cats almost invariably investigate novel objects by sniffing, and communicate by a variety of olfactory signals (see Chapter 5), and the size of the olfactory bulbs and the olfactory membrane, as well as the structure of the olfactory receptors, all argue in favour of a considerable reliance on the sense of smell.

The trigeminal olfactory system

In many mammals, including the cat, it is known that the ethmoid and palatine/nasopalatine branches of the trigeminal nerve innervate the nasal cavity. The receptors are free nerve endings akin to pain receptors, which respond to noxious substances, as well as some odourants, and can trigger protective reflexes. In addition to this protective function, the receptors are thought to show some degree of specificity, since not all have identical patterns of response to different odourants. This raises the possibility that the sensations produced by many odours are in fact due to the combined output of both olfactory and trigeminal systems (Keverne *et al.*, 1986), although this has not been confirmed for the cat.

The vomeronasal organ

The vomeronasal, or Jacobson's, organ occurs in many mammalian species, but not in higher primates, including man. Their structure suggests that they are used only intermittently, as accessory olfactory organs. They are connected to both the oral and nasal cavities by the nasopalatine canal, which runs through the incisival foramen; the lower opening can be seen as a slit immediately behind the upper incisor teeth. The paired vomeronasal organs are blind sacs running backwards from the canal, to which they are connected by very fine ducts, only 30–40 μm wide. Thus the penetration of odours to the chemoreceptors in the organs themselves is unlikely to occur passively, in the way that odour molecules can reach the olfactory epithelium every time that the cat breathes. It has been suggested that a vasometer pumping mechanism expels some of the fluid that fills the sacs out into the canal, and then sucks it back again, drawing in odourants from the mouth and nasal cavity (Eccles, 1982).

The external sign that a cat is using its vomeronasal organ is the gape or 'Flehmen' response, a 'grimace' in which the upper lip is raised and the mouth is held slightly open for a few seconds. This is performed by both males and females, in heterosexual encounters mostly by males, following actual naso-oral contact with urine scent marks or females. Females will respond in the same way to urine marks, if there is no male present (Hart and Leedy, 1987). The need for actual contact with the olfactory material implies that this is a sense more akin to taste than to smell, because the stimuli may be fluid borne throughout. The only stage at which they might be in the vapour phase is in the transfer from the nose and lips to the opening of the nasopalatine canal. The precise role of the gape in sexual behaviour has not been fully investigated, but in other species it has been found that a fully functional vomeronasal organ is essential for successful completion of the first courtship sequence, but that sexually experienced animals can rely on the olfactory sense alone to identify oestrous females.

Conclusion

Many of the differences in sensory abilities between cat and man have been ascribed to the cat's exclusively carnivorous lifestyle. Its prey may be first detected either visually or by sound. The cat's vision functions at much lower light intensities than our own, and is also more highly tuned to the detection of rapid movements. In order to achieve these specializations, the cat has sacrificed a certain degree of visual detail, and the ability to discriminate between the thousands of colours that we can tell apart. The ears of the cat are very sensitive to frequencies that we regard as ultrasound, and it can hardly be a coincidence that these are the frequencies that are used by small rodents for communication. In almost all other respects, man has the more discriminating hearing, probably associated with our need to distinguish between the subtle inflexions of speech.

The approach to prey is aided by the cat's superb sense of balance. Although vision is probably the main sense guiding the pounce, once the cat is actually in contact with its prey its defective close focusing will rule out vision for guiding the kill. At this stage the dominant sense is likely to be that of touch, for which the whiskers, pushed forward in the pounce, the face and the forepaws are particularly sensitive. The integration of these receptors with the killing bite will be described in the next chapter.

The only sense organ which can be categorically associated with social behaviour is the vomeronasal apparatus. Flehmen, the outward sign that the vomeronasal organ is being used, is never seen in connection with feeding behaviour. Olfaction, which is much more sensitive in the cat than in the man, is used when the cat is deciding whether or not to eat, but also has an important role to play in social interactions. It is still by no means certain why cats, and many other mammals, have these two olfactory systems. Part of the reason must lie in the different ways in which the odorous molecules reach the receptors; passively and through a thin barrier of mucus for nasal olfaction, actively and through a thicker barrier (which presumably slows both the onset and decay of the olfactory signal) for vomeronasal olfaction.

The emphasis in the current literature on the cat's adaptations to hunting may simply reflect the relatively recent recognition that it can be a highly social animal as well. As we come to understand more about social signalling between cats, special sensory abilities suited to that purpose may become apparent.

3

Mechanisms of Behaviour

The sense organs described in the last chapter bombard the cat with data at every waking moment. Somehow the important information must be filtered from the irrelevant; this process may start in the sense organs themselves, as illustrated by pattern recognition within the visual system. The translation of sensory input into what we see as behaviour can occur in several ways, and at different levels of complexity. If there is a direct connection between the sensory information and the behaviour pattern, as is the case in simple reflexes, the reaction time will be short, but there will be little scope for flexibility in the response. If the information is thoroughly processed by the brain before any behaviour pattern is triggered, reaction time is likely to be longer, but the stimulus–response relationship can be considerably modified by learning. Some behaviour patterns are so essential for second-to-second survival that a cat could not afford to learn them from scratch; others have to be learned because the relevant sensory information is different for every individual, for example the learning of routes around a home range. It is important for us to try to understand the mental, as well as the sensory, capabilities of the cat, to comprehend fully the subjective aspects of the world that the cat lives in. Every cat owner has ideas of how 'clever' their animal is, but these are usually built on human concepts, some of which are appropriate to a carnivore, and some of which are not.

This chapter deals with the role of the central nervous system and related physiological mechanisms in determining the behaviour patterns that we observe. Underlying rhythmical processes will be dealt with first, followed by the species-specific, reflexive patterns that confer some of the cat's special abilities. Finally, more complex learning will be discussed, again emphasizing those abilities and constraints which seem to separate the cat from other mammals.

44

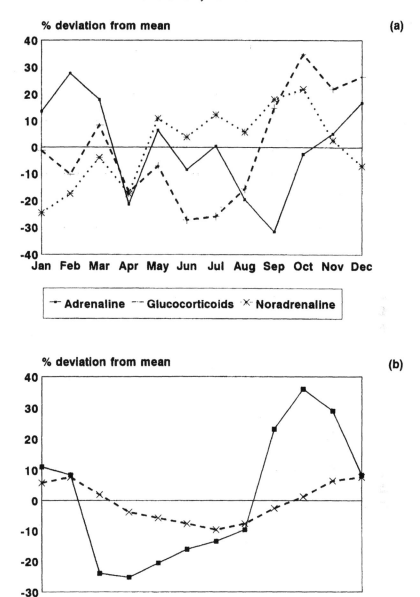

Fig. 3.1. Monthly changes in the levels of adrenaline, glucocorticoids (11-OH) and noradrenaline (a), caloric intake and body weight (b), in cats maintained at constant temperatures (N=8). (Redrawn from Randall and Parsons, 1987.)

Rhythms of Behaviour

Cat behaviour is influenced by underlying rhythms in the endocrine and nervous systems, which are themselves affected by external patterns, such as night and day, and seasonal changes in day length. The annual rhythms have not been studied in great detail (Randall and Parsons, 1987), but the hormones adrenaline and noradrenaline, and adrenaline precursors, the 11-hydroxycorticoids, vary considerably with the seasons (Fig. 3.1(a)). At constant temperature, food intake peaks in the autumn, and is lowest in the spring, while body weight is lowest in the summer and is highest in mid-winter (Fig. 3.1(b)). This relationship between intake and weight is not a direct one, suggesting that metabolic rate may also be subject to annual rhythms. Cats are also affected by daily (circadian) rhythms of activity that are endogenously longer than the normal day, at about 24.2–25 hours (Johnson *et al.*, 1983), but are reset each day by the cycle of light and dark, so that in practice they repeat every 24 hours.

Sleep

The cycle of sleeping and waking is very variable, but is almost always less than 24 hours long, because cats tend to sleep for several short periods during both day and night, rather than in a single sustained session. Sleep has been studied extensively in the cat, and a particular area of the brain-stem, the reticular formation, is known to be a major control centre. Nerve impulses from the reticular formation to the cortex keep the cat awake; these impulses are stimulated in turn by sensory input, both direct, from the sense organs, and also via the cortex in the case of learned signals, such as the visual characteristics of an enemy. There are other interactions; for example, hunger and thirst tend to suppress sleep, acting mainly through the hypothalamus.

The rhythmic patterns in the brain during sleep can be recorded from the skin on the head. When a cat is awake, these electroencephalogram (EEG) patterns have low amplitude but high frequency, and vary greatly depending on how active the cat is, and what it is doing. The onset of sleep is marked by a change to a high-amplitude but much lower frequency EEG, with occasional bursts of medium amplitude, intermediate frequency waves. The cat then looks as if it is asleep, but is readily woken. After about 10–30 minutes, the EEG changes again to low-amplitude, high-frequency patterns rather similar to those of wakefulness, but the cat is now difficult to rouse; this apparent anomaly has given rise to the term para-doxical sleep for this phase. After another 10 minutes or so, normal sleep is resumed, and the two types may alternate if the sleeping bout persists. During paradoxical sleep, there is an almost complete loss of muscle tone, although individual muscles may contract suddenly, bilateral eye movements can be observed, and the tail and whiskers may twitch (Oswald, 1962). All of this implies that cats in this state are dreaming, although we can have no direct evidence for this. Certainly paradoxical sleep seems to be more important than normal sleep, because the less sleep a cat takes, the greater is the proportion of the paradoxical phase.

Reflex Behaviour

Because the behaviour of mammals is so easily modified by experience, it is easy to lose sight of the fact that much of their minute-to-minute behaviour is controlled largely by reflexes. Before discussing the more 'intelligent' aspects of cat behaviour, some of the more pre-programmed patterns will be described. Many of these do not fit into the definition of a simple reflex, which is a brief, stereotyped motor output produced by a standardized input to peripheral receptors, acting by way of relatively simple nervous connections. One example of a simple reflex is the scratching response to irritation on a particular point on the skin; some other examples, from kittens, will be considered in the next chapter.

As the study of neurophysiology has become more sophisticated, it has been possible to study stereotyped patterns of behaviour that are controlled by quite complex, interacting nervous connections; for the sake of simplicity, these will be considered as reflexes also. Many characteristic behaviour patterns can be considered as complex reflexes, because they do not require any input from the fore-brain, the part of the brain that processes much of the sensory information, and is responsible for most learned and 'conscious' behaviour. These include most aspects of loco-motion, including walking and climbing, and the characteristic postures for urination and defaecation, as well as the burying of excreta. The latter, for example, can occur with little or no feedback from the senses, as when a cat, after using a small litter tray, performs stereotyped scratching move-ments in thin air around the tray. Some patterns of agonistic behaviour, such as dilation of the pupils, piloerection, hissing, growling, tail lashing and protrusion of the claws are also reflexive, although others, such as arching of the back, biting and striking out with the forelimbs are elicited from the hypothalamus in the fore-brain (see below). Similarly, some components of oestrous behaviour appear to be reflexive, including rubbing, rolling and calling, the oestrous crouch and treading with the hind legs, as well as parts of the after-reaction (Bard and Macht, 1958).

Posture-maintaining reflexes

The information produced by the balance organs has already been described (see Chapter 2); it is relayed both to the cerebellum in the brain (see below) and also directly to some sets of muscles to form reflexes. Of these reflexes, the simplest are those that trigger contractions of the muscles which direct the eyes, because eye movements do not bring about changes in the orientation of the head to the body, and therefore do not in themselves trigger further signals from the balance organs. These vestibulo-ocular reflexes allow the gaze to be fixed while the head is moving slightly. As the head swivels, the direction and extent of the rotation is picked up by the semicircular canals and is translated into an exactly equal and opposite rotation of the eye. More prolonged turning, in which it would be impossible for the eyes to remain fixed on one point, results in an intermittent repositioning of the eyes through repetitive compensatory movements known as nystagmus. These are rhythmic movements of the eyes, consisting of a slow deviation in one direction, matching the turning of the head, followed by a quick return to approximately the original position. This reflex allows for intermittent clear vision, whereas if the eyes were held stationary in the head vision would be partially disrupted for the whole of the turn.

When a cat's attention is drawn to something to one side, its eyes will move first to look at the object, followed quickly by a rotation of the head, which must be accompanied by a counter-rotation of the eyes if the gaze is to remain on the object. This compensation is driven almost entirely by the vestibulo-ocular reflex system. When a cat is moving, similar reflexes allow the gaze to be corrected for the effects of jolts and jerks due to unevenness in the terrain.

The vestibular system measures small, rapid changes in position or angle much more accurately than large, slow movements, and for the latter the simple reflexes described above would result in under- or over-compensation if used alone. The matching of visual and vestibular signs probably goes on all the time in a continuous learning process. To take an analogy from human experience, this adaptability is shown by the process of adjustment to the wearing of strong corrective spectacle lenses. When first worn, such lenses produce an apparently disturbed motion of the surroundings at the periphery of vision, due to a mismatch between the vestibular signals and the altered visual field, but within a few days these aberrations disappear.

The reflexes in the neck muscles are essentially an error-correcting system. Any rotational displacements of the head will cause the appropriate neck muscles to be activated, such that the disturbing force is counteracted and the head is restabilized. Since any movement of the neck muscles is likely to cause the head itself to move, triggering more signals from the semicircular canals, the detailed working of these reflexes is more complex than those involving the eyes. More complex still are those that trigger contractions of the body muscles, whose effects on the orientation of the skull are unpredictable. One of the simplest of these occurs at the

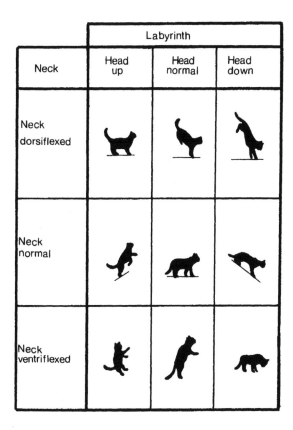

Fig. 3.2. Interactions between the static labyrinthine reflexes and the neck reflexes, and their effects on the limbs. The central figure shows the normal resting posture. In the middle row (left and right) the labyrinthine reflexes operate alone; in the centre column (above and below) the neck reflexes operate alone. Their interactions are indicated in the four corner figures. See text for further interpretation. (Redrawn from Wilson and Melville Jones, 1979.) (M.T.)

beginning of a fall; within 70 ms of a cat losing its footing, signals from the otolith balance organs trigger extensions of the legs, as a preparation for landing (Watt, 1976). The semicircular canals stimulate reflexes that will tend to restore body position; for example, if the head rotates to the left, both front and hind left legs are extended, while both right legs are flexed. However, in many real instances the neck reflexes will act first, and thereby complicate the extent and direction of the body reflexes.

These interactions are made more complicated by reflexes, acting on the limbs, that are triggered by the position of the neck itself, independently of the reflexes originating in the vestibular organ. Some interactions between the static (otolith) reflexes and reflexes originating in the neck are illustrated in Fig. 3.2. Together these act to maintain body posture, so that movements of the head do not trigger inappropriate movements of the limbs. For example, if the cat raises its head to look up (top left of Fig. 3.2), the otolith reflex will tend to flex the front legs and extend the back legs to restore the head to the horizontal. However, this tendency is cancelled out by the neck reflexes, which are acting simultaneously to extend the front legs and flex the back legs. The net result is that the cat is able to raise its head without either front or back legs moving involuntarily. The exactly opposite reflexes, which also cancel one another out, allow the head to be lowered independently of movements of the legs (bottom right of Fig. 3.2). Other examples in this figure show how head and limb positions interact when the cat is jumping up or down from one surface to another (bottom left and top right respectively).

Locomotion

The basic patterns of locomotion are contained in spinal 'programmes' which produce the main features of rhythmic stepping for the various gaits described in Chapter 1. The spinal cord contains specialized autonomous stepping generators for the hindlimbs, and probably also for the forelimbs. Each of these contains the pattern for a complete step of a single limb, which can be speeded up or slowed down as necessary. Alternative neural pathways between these generators allow for the different ways in which the individual limbs follow one another to produce the various gaits. Signals from proprioceptors in the limbs allow for corrections due to, for example, uneven terrain, while the whole programme is activated and sustained by command signals from the brain (Wetzel and Stuart, 1976).

The orienting reflex

Cats, like most mammals including man, will rapidly orientate their sense organs towards any sudden event in the environment. This involves complex movements which are situation specific and therefore far from stereotyped, so the term reflex is being used here in its sense of there being no involvement of conscious thought in the behaviour pattern. The orienting reflex was first characterized by Pavlov, who discovered that the motor patterns involved were not specific to either the quality or intensity of the stimulus, which can be provided by any one of the senses, or a combination. The most important feature of the stimulus is its novelty; the ending of a continuous stimulus such as a drawn-out sound evokes the same response as the beginning of that sound. If the same stimulus is repeated over and over again, the reflex becomes weaker, and is finally not elicited at all. In the brain, one major effect is the dilation of the cerebral blood vessels and constriction of the peripheral, which facilitates the transmission of information through the central nervous system, making the cat more 'attentive' (Sokolov, 1963). The essentially pre-programmed nature of this reflex can be illustrated by its invariant effects on the eyes. If an object appears suddenly in the visual field, the pupils dilate, and the eyes automatically focus at their shortest possible distance, even if the object is actually far away. Non-visual stimuli have exactly the same effect on the eyes, whether they are odours, sounds or a light touch, always provided they occur with an element of surprise.

The Brain and the Control of Behaviour

The brain, and particular the fore-brain, exerts a controlling influence at almost every stage of the more complex reflexes. For example, the reticular formation in the brain-stem not only controls sleep, but also the general state of arousal. It influences the impact of all the sensory systems on the cerebral cortex, and is particularly active during habituation, the process whereby the same stimulus, if repeated, elicits a weaker and weaker behavioural response. A second, parallel arousal system in the mid-brain mediates the effects of learned behaviour patterns (Colgan, 1989).

In addition to such non-specific effects, it has proved possible to group some behaviour patterns together, based upon the site in the brain from which they originate. One of the best understood is the 'quiet biting attack', which is the psychologists' term for the patterns seen in the latter stages of hunting, culminating in the kill. Groups of neurones in the hypothalamus and mid-brain control a whole sequence of events, each one of which contains several reflexes. In the order in which they occur, these are:

1. stalking, sniffing, and visually guided approach to the prey;
2. visually guided orientation of the cat's head to the target, assisted by tactile stimulation from the forepaw if this makes contact with the prey;
3. when the head reaches the target, precise orientation of the snout by tactile stimulation of a trigger zone on the face;
4. opening of the jaws, in response to stimulation of a trigger zone around the lips;
5. finally, closure of the jaws when a trigger zone just inside the mouth is activated.

The hypothalamus has an important role to play in changing the thresholds for the component reflexes. For example, the seizing and biting reflexes are switched on, while others which would interfere with the capture of prey, such as the jawdrop reflex, are suppressed. The sensory inputs required at each stage can be defined precisely; the head-orienting response occurs in response to touch over an area of skin extending from just above the upper

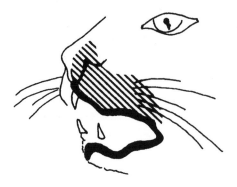

Fig. 3.3. Sensory fields which trigger the head-orienting (shaded area) and jaw-opening (solid areas) responses during prey capture. The lower jaw also contains a less well-defined area (not shown) directing the head-orienting response. (Redrawn from Macdonnell and Flynn, 1966.) (M.T.)

lip to the hairless area on the nose, and out to the side as far as the whiskers. Biting requires a touch on either the upper or lower lips, most effectively at the front of the mouth, but to a lesser extent around to the sides (Fig. 3.3). Persistent biting requires stimulation of the trigeminal receptors around the mouth, as well as touch receptors (Siegel and Pott, 1988).

Quite separate areas of the hypothalamus and mid-brain control a group of defensive behaviour patterns, including retraction of the ears, pilo-erection, arching the back, marked dilation of the pupils, vocalizations and unsheathing of the claws. Yet other areas of the brain control flight behaviour. Thus the way that many species-specific patterns are organized in the brain mirrors the groups that we can place them in, based on their functions in free-ranging animals.

Comparison with other species

We can also deduce something of the special features of the cat's brain by making comparisons with other species. One concept that has proved successful in such comparisons is that of structural encephalization, which is defined as the enlargement of the brain beyond that expected from the size of the body, and is measured as an encephalization quotient (EQ) (Jerison, 1985). Large bodies need large brains because they have larger muscles and more extensive somatic sensory systems, but once this is allowed for some striking comparisons can be made. For example, deer, wolves, crows and lemurs all have roughly the same EQ, while hedgehogs have retained the lower EQ of the earliest mammals. This measure contains an element that could be defined as 'intelligence', but enlargement of the brain as a whole can also be due to specializations, such as an increase in the sensitivity of one or more of the senses, which will produce more information for processing. Some of the ecological factors that have been proposed as requiring an increase in EQ are: movement in three dimensions (flying, swimming) compared to two (terrestrial); an active antipredator strategy compared to a passive one (e.g. the hedgehog); and a long period of parent–young association for the transfer of skills (Shettleworth, 1984). More nebulous, but intuitively correct, is the idea that learning abilities differ between animals in terms of how flexible that learning can be. Higher primates can learn a wide variety of tasks and associations; the learning abilities of lower mammals seem to be more situation specific, more constrained by the niche for which they have evolved. The former is likely to raise the EQ more than the latter. Thus EQ is built up from several components, which can have very different emphases in different species; overall the figure reflects an investment in information-processing power, whether it be for learning, or for a special skill, or for sensory ability. The first of these is the most flexible, the latter two are likely to be more niche-specific.

The brain of the domestic cat is very similar to that of other members of the genus *Felis*. The basic pattern appears in the fossil record some five to nine million years ago; the brains of earlier cats, most of them sabre toothed (the paleofelids), appear to have been organized along different lines (Radinsky, 1975). Two of the most striking features of the brain of the domestic cat are the enlargement of the cerebellum, co-ordinating balance and movement, and the large proportion of the cortex devoted to controlling movement; cats' brains reflect their athletic prowess. The part of the cortex that deals with hearing is well developed, but the olfactory bulbs are, compared to other carnivores, rather small. The felids as a family have rather little space for large olfactory bulbs in their comparatively short skulls; those that do have large olfactory bulbs, like the lion, have large home ranges, a trend repeated across all the Carnivora (Gittleman, 1991). The visual area of the cortex is less developed in the domestic cat than in some other fields, for example the jaguarundi, *F. yagouaroundi.*

The cat's EQ is higher than for the majority of rodents, and about average for the carnivores. The dog family has the highest average EQ of any carnivore, some 25% higher than for the average felid; larger olfactory bulbs in dogs, reflecting a greater reliance on their sense of smell, are partly responsible, but the prefrontal cortex, which is thought to selectively inhibit primitive behaviour patterns, is also larger in some dogs. It has been suggested that such inhibition, for example the substitution of aggressive by submissive behaviour patterns, may be a component of the complex social behaviour seen in wolves and other canids.

Learning and Intelligence

It is impossible to draw a sharp distinction between instinctive and learned behaviour in an animal as complex as the cat. Species-specific behaviour, such as vocalization, mating behaviour, some aspects of hunting, and the reflexes displayed by kittens, are presumably based upon inherited patterns, but these are modified, supplemented and altered, in both the long and the short term, by learned components. Although the cat has been a favourite subject for the study of learning itself for the past 100 years or so, many of these studies have taken little account of the evolutionary pressures that have shaped the mind of the cat, compared to those that shaped, for example, the rat or chimpanzee. Such considerations of the unique character of every species have led to a recent trend away from comparing their mental abilities, on the grounds that each has been shaped to fit it for its own niche, and so there is no logical basis for such comparisons. On the other hand, it is self-evident that an ape has more mental capacity than a mouse, even allowing for the anthropomorphic tendency to classify human abilities as 'intelligent'. Biological and psychological approaches have recently been brought together in the concept of 'ecologically surplus abilities' (see Davey, 1989). This argues that while the vast majority of the species are closely adapted to their current niches, their abilities to respond to sudden changes in those niches vary considerably. Animals relying largely upon instinct, or highly context-specific learning, will only be able to readapt at a pace determined by evolutionary mechanisms. Those with more extensive learning abilities can alter their behaviour patterns rapidly. Thus 'ecologically surplus ability' can be defined as the capacity to solve problems which have not been selected specifically by adaptation to current niches, but are available to cope with unexpected change.

The domestic cat seems to be a prime example of a species with ecologically surplus abilities, given that it is able to move from total dependence on man to semi-independence and back, within a very few generations. Such abilities are, almost by definition, difficult to assess, since their full value will only be expressed under circumstances of rapid change in the environment. However, certain mental skills, such as learning by imitation, and the formation of mental concepts, are likely to contribute to the flexibility required, and these are described in the section below on 'Complex learning'. On the other hand, cats are not infinitely flexible, for there is ample evidence that their learning abilities are species specific at all levels of complexity, including straightforward associative learning.

Learning by association

At its simplest level, learning involves the linking together of previously unrelated stimuli, or actions and the consequences of those actions. Many invertebrate animals are capable of this type of learning, and so it is hardly surprising that cats can form a wide range of associations of this type. Indeed, in the past the behaviourist school of animal psychologists has attempted to describe all learning at this level, but it is now thought that many mammals, including cats, are capable of much more complex mental processes, some of which will be described in the next section. By contrast, popular accounts of cat behaviour tend to express learning in the terms of human subjective experience, almost as if cats were mentally defective people rather than highly adapted carnivores.

Such controversies are far from new, as the following excerpt from Hobhouse (1915) will illustrate:

> I once had a cat which learned to 'knock at the door' by lifting the mat outside and letting it fall. The common account of this proceeding would be that the cat did it in order to get in. It assumes the cat's action to be determined by its end. Is the common account wrong? Let us test it by trying explanations found on the more primitive operations of experience. First, then, can we explain the cat's action by the association of ideas? The obvious difficulty here is to find the idea or perception which sets the process going. The sight of a door or a mat was not, so far as I am aware, associated in the cat's experience with the action which it performed until it had performed it. If there were association, it must be said to work retrogressively. The cat associates the idea of getting in with that of someone coming to the door, and this again with the making of a sound to attract attention, and so forth.... Such a series of associations so well adjusted means in reality a set of related elements grasped by the animal and used to determine its action. Ideas of 'persons', 'opening doors', 'attracting attention' and so forth, would have no effect unless attached to the existing circumstances. If the cat has such abstract ideas at all, she must have something more – namely, the power of applying them to present perception. The 'ideas' of calling attention and dropping the mat must somehow be brought together. Further, if the process is one of association, it is a strange coincidence that the right associates are chosen. If the cat began on a string of associations starting from the people in the room, she might as easily go on to dwell on the pleasures of getting in, of how she would coax a morsel of fish from one or a saucerful of cream from another, and to spend her time in idle reverie. But she avoids these associations, and selects those suited to her purpose. In short, we find signs on the one hand of the application of ideas, on the other of selection. Both of these features indicate a higher stage than that of sheer association.

Hobhouse evidently interprets his cat's behaviour as having purposeful elements. However, he does go on to offer an alternative explanation, which we would now class as behaviourist. This is based upon an association between the 'pleasure' of getting through the door, and the action of lifting and dropping the mat, no more. The action assimilates the character of its result and becomes in itself attractive to the cat.

Pavlovian learning

Even with such a simple sequence of events as this, psychologists do not agree on the precise details of the learning mechanisms involved. Their findings, usually based on simple tasks carried out under highly defined conditions, are not always easily interpreted in functional terms, i.e. the value to the cat of the type of learning observed. One of the simplest forms of learning is known as Pavlovian conditioning, named after the classic experiments of Pavlov, who trained dogs to anticipate the arrival of food in response to arbitrary signals, such as the sound of a bell or metronome. The dog has continued to be a favourite subject for this type of study, so some of the examples to be described, although drawn from the dog, will be assumed to apply to the cat also. The primary function of Pavlovian learning seems to be the acquisition of information about stimulus relationships in the environment. One stimulus, the Unconditioned Stimulus (UCS), is normally linked to a particular motivational state, and releases an innate reaction, the Unconditioned Response (UCR); for example, the smell of food will result in salivation in a hungry animal. If a second stimulus, the Conditioned Stimulus (CS), occurs consistently with or immediately before the UCS, it will come to release the UCR even on its own; the UCR has become a Conditioned Response (CR). However, the UCR and CR need not be identical, although they are normally linked to the same type of motivation. For example, if the UCS is the pain inflicted in an attack by an aggressive tom cat, the UCR will most likely be flight. The victim will probably associate the appearance of the attacker (the CS) with the experience of the fight, and will respond appropriately (CR) on seeing the attacker again. However, this response may change with circumstances; if the CS is fairly distant, the CR may be to 'freeze' in an attempt to avoid detection, rather than to run away, inviting a chase. Such considerations, and more direct experiments, have led to the conclusion that in most cases the association built up in Pavlovian conditioning is genuinely formed between the CS and the UCS, and not between the CS and the CR; in ethological terms, if the UCS is a releaser, the CS is a learned releaser, bringing it under the control of normal motivational systems.

The interspecific differences that this can produce have been illustrated by a comparison between the learned feeding responses of cats and rats (Grastyan and Vereczkei, 1974). The arrival of a food reward was signalled by 10 s of a clicking sound, coming from a loudspeaker 2 m away from the food magazine. This combination caused the cats to run towards the sound, and some would search all around the loudspeaker, and even attempt to bite it. When this response was most intense, the cat would often not take the food reward at all, although after hundreds of further trials the feeding response was re-established. Under similar conditions rats would briefly turn their heads towards the sound, but would rarely approach it. For the rats, the sound was an initially irrelevant cue, but for the cats, which use auditory cues extensively while hunting, it was not, and evidently some conflict appeared between the apparent location of the 'prey' as indicated by the sound, and its subsequent appearance as food.

An important feature of the relationship between CS and UCS is that they must be highly correlated; if the pairing is unreliable, the response (CR) is considerably weaker than when the CS and UCS always occur together. This prevents a cat from acquiring false or poorly predictive information about its environment. Events which reliably do not predict the UCS are also learned, as can be shown in two separate ways. First, if a CS is repeatedly presented in a way that does not predict the arrival of the UCS, then when the same CS is presented with the UCS it is difficult to establish the connection; the cat has already learned that the CS is an irrelevant cue, and so when its prediction value changes there is a delay before the new association is registered. Secondly, if the procedure described is performed in reverse, the association is rapidly 'unlearned' once it is no longer predictive. Furthermore, associations can be learned between two neutral stimuli (i.e. not releasers) which reliably occur together, even when neither brings about any overt behaviour. This can be shown by pairing just one of the two stimuli with a UCS, after which both stimuli will release the CR. This 'behaviourally silent learning' is of obvious value to, for example, a cat learning the topography of its home range, although cats also possess more advanced orientational abilities, which will be described in a later section.

Pavlovian learning is probably the basic mechanism behind many other behavioural phenomena, including taste aversion learning (discussed in Chapter 6), and some aspects of foraging behaviour. For example, if cats form the equivalent of the 'search images' used by birds to detect cryptic prey (but see Guilford and Dawkins, 1987), they may do so by associations between the appearance of a specific prey type and its profitability. Specific features of the environment may come to be associated with particular prey types or prey densities. Moreover, Pavlovian associations can also prepare the cat for subsequent events so that they can be optimally exploited; for example, CSs indicating food bring about physiological changes that speed up digestion once the food is actually eaten.

Instrumental learning

These simple Pavlovian mechanisms should enable a cat to build up a much more organized picture of its world than would instinct alone, but they will not on their own produce the flexibility in behaviour that cats are evidently capable of. For the latter, a different type of learning is required, one that will enable the cat to predict the consequences of its own actions, and modify those actions based on past successes and failures. This is addressed by the psychological technique of instrumental learning, in which the subject has to respond in some way to a stimulus; correct responses are rewarded. Some of the earliest work in this area used cats as subjects, particularly the puzzle-box experiments of Thorndike (discussed in Hobhouse, 1915). Thorndike placed cats in cages from which they could escape by means well within their motor capabilities, such as clawing at a string, depressing a lever, pushing aside a swing door, and so on (Fig. 3.4). The cats would claw and scratch indiscriminately at the sides of the cage, until by accident they performed the right action, and gained their freedom. The time that it took the cats to escape declined with repetition, implying that the probability of performing the correct action was increased by each success. Some of the tasks set were quite complex; one latch required a simultaneous lift and push, and in other cages two or even three latches had to be opened in the correct sequence. Although not all cats could master these, all were opened by some. Taking an average of several animals, the skills appeared to be gained gradually, and Thorndike concluded that 'The gradual slope of the time-curve, then, shows the absence of reasoning. They represent the wearing smooth of a path in the brain, not the decisions of a rational consciousness.' However, individual animals did not behave quite in this probabilistic way. Some did take a little less time to escape on each successive attempt, but many seemed to

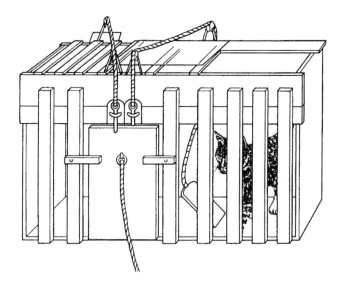

Fig. 3.4. An example of one of Thorndike's puzzle boxes. (From McFarland, 1985.)

improve their performance quite abruptly, and then never make another mistake, even with an interval of several months between trials. In fact, rapid (one-trial) learning is nowadays not thought to be good evidence for conscious thought. Many animals learn crucial associations, such as the toxicity of particular foods, after only one experience; in this situation the potentially lethal consequences of eating the same food again are likely to outweigh the risk that, after only one encounter, the animal has learned the wrong connection.

Thorndike's idea that random behaviour patterns were shaped by successes gave this type of process its alternative name, 'trial-and-error learning'. The apparently random behaviour of the cats when first put into the cages, together with the results of many other such experiments on other species, helped to establish the notion that almost any behaviour could be shaped in this way, minimizing the value of instinctive behaviour. However, it is now clear that species-specific behaviour patterns have a clear role to play in providing the behaviour that is to be shaped, in directing the attention of the cat towards the task to be performed, and in providing an assessment of the value of the reward for correct performance. These species-specific constraints presumably ensure that in the real world, outside the narrow context of the typical instrumental learning paradigm, the most ecologically functional skills are acquired. Thus it is much easier to train a cat to obtain a food reward by using a normal component of hunting behaviour, such as hooking back a bolt with its paw (the movement used to dislodge prey that takes refuge in a crevice), than by some arbitrary but straightforward action, such as pushing an identical bolt inwards. In the cage experiments, Thorndike found that certain actions of the cat could not be trained; for example, if the cat dislodged the latch by accident with its tail, it did not appear to learn anything about the location of the latch or type of actions likely to open it on subsequent trials. Also, if the cat was allowed to escape every time it performed some arbitrary action, such as grooming, the frequency of grooming did not increase; the connection between the action (grooming) and the reward (escape) was never made. The type of incentive is also important. For young cats, although food is a powerful reinforcer, other activities, such as manipulation of simple objects such as a ball or a crumpled piece of paper, or exploration of an unfamiliar space, are also adequate rewards for a discrimination task (Miles, 1958).

Instrumental learning methods have been used extensively to probe the sensory and mental capabilities of cats; examples of the former have been described in Chapter 2, and the latter will be explored in more detail in the following section. To provide information on the way that the cat makes everyday decisions about its actions, more complex schedules of reward and response are required. For example, one common procedure is to reward two distinct responses simultaneously, either pairing each with a reward of different value, or rewarding different proportions of the two responses. In many species, the strategy that is adopted can depend on the type of stimulus (e.g. visual or spatial) used to indicate the reward, presumably reflecting species-specific propensities to learn links between food and its sensory characteristics. The types of strategy adopted by the animal can give some idea of the way that they might behave in the field when confronted with prey items of different nutritive value, or which are more or less easy to catch. One set of trials that mimicked the latter situation (Warren and Beck, 1966) can be used to illustrate the procedure and its possible results. Cats were rewarded intermittently for choosing one of a pair of visually distinct wooden blocks (e.g. one triangular and black, and the other circular and white). If only one block was rewarded (reinforcement ratio 100 : 0), the cats rapidly learned to respond only to that block. If both were rewarded equally (50 : 50), responses were, on average, also equal. A 90 : 10 ratio resulted in all cats responding in the same way as to the 100 : 0, a strategy known as maximizing, because by this means the maximum number of food rewards are obtained within a session. The more complex reinforcement ratios of 60 : 40, 70 : 30 and 80 : 20 produced some very individual-specific shifts in strategy. Particularly at the lower ratios, most cats distributed their responses within 5% of the reinforcement ratio, a strategy known as matching. The ratio at which each cat switched strategies varied considerably, some maximizing at 60 : 40, others matching up to 80 : 20. The mechanism behind matching seemed to be a simple one; most cats persevered in responding to the stimulus which had produced food on the previous attempt, only switching to the other when this prediction failed to pay off. The origins of the individual differences could not be determined, but may have been due to the cats' previous training experiences.

It is still difficult to extrapolate from the results of such trials to real foraging decisions, because they still contain an element of artificiality, in that the cats obtain a great deal of their daily food intake away from the training procedure. More recent closed-economy experiments, in which animals have to do work for all their food, have produced some conclusions that are opposed to those obtained from trials like the one described above (Davey, 1989). Very few trials of this type have used cats, but there is some evidence that the maximizing strategy can be subservient to a direction-stable strategy in which each cat has a preferred foraging route (see Chapter 7).

Various extensions of instrumental learning are used when cats are taught to do tricks (Voith, 1981). It is sometimes claimed that cats cannot be taught tricks, but what is usually meant by this is that cats cannot be taught by the same methods as can dogs. Most dogs are very attentive to their trainers, and can be rewarded by positive social contact alone. Cats are much less likely to be interested in the training process for its own sake, and need to be rewarded with food. The sooner the reward is given after each correct response, the more rapidly will the learning take place. Instrumental learning tests also show that cats, in common with most mammals, remember tasks for much longer if they are only rewarded for a proportion of correct solutions, for example one in five, and this can be used to fix the results of training. These simple techniques can be used to induce cats to perform normal behaviour patterns on cue; patterns which are not entirely natural require a further extension of the technique, known as shaping. To take a simple example, cats will not usually jump over an obstacle if they can walk round it. To train a cat to jump on request, it can first be rewarded for walking over a stick that is lying on the ground, then for stepping over it when it is raised slightly. As the stick is raised further, the cat is only rewarded for jumps. Once the habit is established, it can be made more persistent by only rewarding a proportion of successes.

More complex tricks often have to be built up a step at a time, and for this the technique of conditioned reinforcement is often used. If the 'real' reinforcer, for example food, is always given in conjunction with a signal, such as a sound, then by the time the training for the first component of the trick is completed, the sound will have become a substitute for the food reward, through Pavlovian conditioning. The food can then be used to reinforce the training for the second component, while the sound becomes the sole reward for the first component. If this process is repeated, quite long chains of shaped behaviour patterns can be built up, only the last of which is ever rewarded with food; for example, cats can be taught to retrieve objects in this way.

Complex learning

The simple processes of instrumental learning and Pavlovian conditioning can be demonstrated in invertebrates as well as vertebrates, and so cannot in themselves explain all of the cat's mental abilities. The cat is no longer a favourite subject for the study of learning – much more is known about the specific abilities of pigeons, rats and monkeys – and so the account that follows is by no means a complete description of feline intelligence.

Complex stimuli

Ecologically meaningful cues are rarely simple; they may differ from their background, and other less relevant cues, in several ways, for example size, shape, brightness, colour, characteristic movements, sounds and odours. A great deal is known about the ability of cats to discriminate between stimuli that differ in only one sensory dimension, but much less about the analytical processes which they use when confronted with complex stimuli. Some idea of these processes can be gained from experiments carried out to

Rewarded

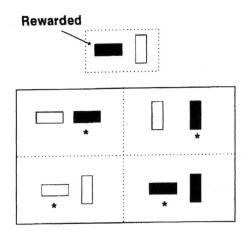

Fig. 3.5. One example from the sets of discriminations that show that cats can learn two attributes of a visual stimulus simultaneously. Young cats were rewarded for responses to the left-hand figure in the upper box, and once trained were tested for preferences between the four pairs in the lower box. Each of these differed in either shading, or orientation, but not both. The cats tended to make choices (starred) indicating that they had learned that the rewarded object had been both shaded and horizontal. (Redrawn from Mumma and Warren, 1968.)

detect the most relevant of a pair of cues presented simultaneously (Mumma and Warren, 1968). Three-month-old cats were trained to distinguish between rectangles that differed in both shape and brightness, and were then tested to see which one they preferred of pairs of rectangles that differed in either shape or brightness (Fig. 3.5). These preferences showed that both cues had been learned simultaneously by most kittens; although there was the expected variability in accuracy, there was no evidence that some had relied more on one than the other, as rats tend to do. The relevance of particular shapes has also been examined, and one cue that cats seem to pay great attention to is whether figures are open or closed. For example, they learn to discriminate circles from U-shapes much more quickly than from triangles (Fig. 3.6). The basis for this seems to be the ratio between the area of a shape and the number of sides that it has (Warren, 1972). When irrelevant cues are presented along with relevant ones, cats are better than rodents at singling out the predictive one. For example, from a set of wooden triangles and circles that could be black or white, and one of two sizes, only triangular or circular shapes were rewarded with food. The cats learned that it was the shape that was the discriminating feature, and there was no difference in the speed of learning between cats trained with pairs of triangles and circles which were always the same shade and size, and cats trained with pairs of triangles and circles the shade and shape of which changed from session to session (Warren, 1976). The errors that cats make in such trials seem to originate in their initial preferences and aversions, which they rarely overcome as completely as monkeys do when trained on the same problems.

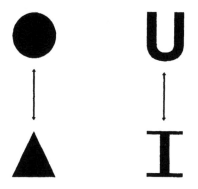

Fig. 3.6. Examples of shapes that cats find easy to discriminate from one another, the exceptions being the pairs connected by arrows.

The concept of oddity

The precise extent to which cats can generalize from one discrimination to another is still unclear. One such generalization is that of oddity. Chimpanzees can quickly grasp the idea that they are to pick out the non-matching object in a group of three in which the other two objects are identical. Cats take much longer to learn this, and are prone to mistakes. In one set of trials, five nine-month-old cats were initially trained to discriminate the odd one (for example, a triangle) of three objects (the other two being, for example, circles), when the same set of three objects was presented at each session (Warren, 1960). Once each cat had successfully learned that it should look for the triangle, one of the circles was removed from the set, and was replaced by a triangle; in other words, the oddness was transferred from the triangle to the circle. The odd object was still the one that was rewarded; initially the cats preferred either of the two triangles, because that shape had previously been associated with food, but quickly turned their attention to the circle, which was now the odd object. This reversal was repeated 20 times, and each time the cats followed the change, although one of the five was consistently more accurate than the others. This part of the procedure demonstrates that the cats could discriminate the objects from one another even when two were identical, but it does not demonstrate learning of oddity itself. In the second part of the procedure, the triangle/two circles and circle/two triangles combinations were presented in a random order, and one cat, the best performer in the preliminary trial, learned that it was the odd one out, rather than either of the shapes themselves, that signalled food. This cat could also rapidly generalize from this pair of shapes to others; presented with random orders of two new shapes in groups of three, it made fewer and fewer errors each time the pair of shapes was changed for a new pair. That the cat that mastered the oddity problems was also the best at recognizing objects, suggests that these two types of learning are linked. It is also possible that all the cats had the concept of oddity, but could not be persuaded to demonstrate it by object discrimination.

Object permanence

Similar problems may lie behind disagreements about the extent to which cats understand where objects have been hidden. Such skills would be highly adaptive for a carnivore hunting in cover, and so we should expect cats to be highly aware of the most likely location of prey that has gone to ground. A theoretical framework for the concept of object permanence, devised by Piaget for recording the development of human infants, has been used to quantify the abilities of cats and is therefore worth summarizing in its original form. In the first two stages of development, infants show little interest in objects, and when an object is hidden, they stare at the point from which it disappeared, rather than looking round for it. Stage 3 is marked by the ability to discriminate partly hidden objects, and to recognize the part that is visible as belonging to the original whole. Stage 4 introduces the concept of permanence for the first time; objects that disappear are searched for, indicating that the infant realizes that they still exist. However, in a series of tests with the same object they tend to search the place where the object has been hidden most often, rather than the place where they have just seen it hidden; a previously successful action is repeated, akin to the result expected from instrumental conditioning. Reliance on immediate perception is established in Stage 5, and more complex problems can also be solved. These include sequential visible displacement, in which an object is hidden in several places in turn, the solution being to look in the place closest to where it was last seen, and single invisible displacement, in which an object is hidden first in the hand, and then the hand is placed under a cover. When the hand is withdrawn and shown to be empty, the child should look under the cover. The final (6th) stage completes the mental concept of object permanence, in which the child can follow sequential and successive invisible displacements. In the first, an object is hidden in the hand, put under one cover, shown to be under that cover, palmed again, and hidden under a second cover. Infants at Stage 5 tend to look under the first cover. In the successive displacement, the object hidden in the hand is moved from one cover to another, and left under the last before the empty hand is displayed; again, infants that have not reached this stage tend to start by looking under the first cover.

The extent to which cats can be persuaded to demonstrate their abilities in this area seems to depend a great deal on the type of object that is used. Cats seem to be unable to solve simple invisible displacement problems when the object is a toy, but this may be because the apparatus used to hide the toy is as interesting to the cat as the toy itself; it 'knows' that the toy is under a cloth, but has no inclination to find it. When the object is food, however, cats pay much more attention and are able to demonstrate their abilities much more readily. Thus cats can quickly solve the successive invisible displacement problem, which is the last to be solved by human infants. The cat is allowed to watch the food being hidden in a cup, and the cup is then hidden in turn under three covers, after which the empty cup is shown to the cat. To control for odour cues, the food is not actually deposited under the last cover. Triana and Pasnak (1981) describe one of their cats' first reaction to this problem thus:

> As soon as the cup was removed from under the left cover and shown to be empty, the cat hurried to this cover (not to the experimenter's hand where the food, in fact, was). The cat then persistently pushed back the cover until the place where the food should have been was entirely revealed. The cat continued to poke at the cover with its forepaws and tried to push its face underneath for approximately 3 more minutes.

However, this cat had already learned that covers were likely hiding places, and so may have been repeating a previously successful strategy. Other studies have failed to show any ability for invisible displacements. and have concluded that cats look for objects in the place where they disappeared, or under the nearest cover to that place. In terms of hunting behaviour, this means that if prey becomes temporarily invisible, a cat will search for it under or behind the place where it disappeared, and then if unsuccessful will probe the nearest available cover. Familiarity with the environment will be important to enable the cat to know the most likely hiding places, and therefore if the environment is not familiar, the chances are reduced of the cat finding prey with which it has lost visual, auditory and olfactory contact.

Concepts for orientation

Familiarity with the environment implies that cats have some kind of concept of the way that the components of the world around them fit together. This has been investigated by examining the ways that cats find their way around. There are several possibilities, incorporating different levels of sophistication. The simplest type of orientation relies on direct perception of the goals ('the rabbit warren is in the bank that I can see at the other side of the field'), or a step-by-step route based on landmarks ('if I go to the oak tree that I can see, and turn left, I will then be able to see the warren'). Many animals, including some invertebrates, use such orientation systems, which are generally simple to use, but prone to error ('the oak tree has been felled, so I can no longer find the warren'). Cats may rely on these systems in simple situations where they are unlikely to lead to error, but they are also capable of constructing cognitive maps of their surroundings, particularly if they have been able to explore them thoroughly (Poucet, 1985). Although they can also construct maps based on a brief view of relevant features, these are not remembered for more than a few minutes. Mapping leads to the possibility of taking short-cuts ('Last time I went to the warren I went to the oak tree and turned left, so this time I will go diagonally across the field and through the hedge; the warren is just beyond the hedge'). It also permits the rapid choice of optimum routes; given a choice of ways to an invisible goal, cats almost always prefer the shortest one. If there are several routes of roughly the same length, the one that starts off in the direction closest to the direction of the goal itself may be preferred, a common human habit also. Minimizing the number of twists and turns in the route after that is also a factor that determines a cat's choice, but a relatively unimportant one.

Cats are also capable of discriminating short time-intervals; they can tell the difference between a sound that lasts four seconds from one that lasts five, and can also learn to delay their response to a stimulus by several seconds, again to an accuracy of about one second. This implies the existence of an internal clock which times the duration of both internal and external events; this could be used, for example, in assessing the rate at which particular feeding strategies produce food. Another skill which would be useful in this context is the ability to count, and it is thought that cats do have some kind of abstract conception of number, although attempts to demonstrate this have not proved recognition of numbers greater than about seven.

Finally, it is worth returning to the question of how cats obtain the information on which learning is based. Trial-and-error is a time-consuming process, and in a social animal a great deal of time could be saved by watching the ways that conspecifics solve problems. Cats are certainly capable of this, even when they are adult (John *et al.*, 1968), although it has been argued that the actions of the conspecific merely help to focus the cat's attention on the problem to be solved. Some are apparently able to 'work out' exactly how to perform a task simply by watching an experienced individual carry out that task, and then repeating the actions they have seen. Learning of this kind is essential in the most intense period of the cat's social life, its life as a kitten with its mother and siblings.

Behavioural Development

There is a sense in which the behaviour of a mammal is always developing, because learning is a process that starts in the womb and persists into old age. However, many of the changes we see in both young and old cats are due to a combination of learning and structural changes that occur in the sense organs and the central nervous system; maturation at the beginning of life, and degeneration at the end. Many aspects of these processes have received intensive study in the cat, although there are still considerable gaps in our knowledge, particularly in relation to the changes that accompany old age.

Maternal Behaviour

For the first few months of their lives, kittens spend a great deal of time with their mothers, and so no account of their behavioural development can be complete without putting it into the context of maternal care. If the mother is a pet, then there will also be a considerable influence from the human caregiver, but consideration of this aspect will be left until Chapter 9. If the mother is part of a social group, there will be considerable input from other females in the group, particularly those that produce their own litters at about the same time; communal rearing of kittens is discussed in Chapter 8. The account that follows is based on the behaviour of solitary females, and of necessity much of the most precise information has been obtained from cats confined indoors, rather than solitary outdoor cats that tend to keep their young litters in inaccessible places. Paternal influences, apart from those conveyed by genetics, are likely to be insignificant, since the pair-bond in cats is usually very weak, and in any case most queens are promiscuous given the opportunity.

A pregnant female will spend a great deal of time and effort seeking out a suitable place to give birth, and may visit and revisit several alternatives before coming to a decision. The site is generally well protected, but there is little or no actual construction of a nest. The selective pressures that shaped these behaviour patterns are not all readily apparent, because they probably evolved under different circumstances to those in which most feral cats find themselves today. Protection from the weather must be an important factor in temperate climates, where kitten mortality due to respiratory and enteritic viruses is high. Potential predators of young kittens include dogs, possibly some birds, and other cats, although the impact of the latter is still controversial (see Chapter 8). After parturition the mother may become aggressive towards cats and dogs that she had previously tolerated. The kittens may be transferred to a new nest site even while they are still helpless; the reasons behind such moves are not fully known, but they may be an attempt to reduce the transmission of parasites, or because the original nest has become fouled with the remains of prey and excreta. Some of these moves bring about the pooling of two or more litters, or occasionally the division of a joint litter, and so may occur for reasons related to social behaviour.

Prior to the birth the queen cleans herself thoroughly, particularly her ventrum around the nipples, and her anogenital area. The residual saliva may leave olfactory cues which help the newborn kittens to locate the nipples. As each kitten is born, she cleans away the birth membranes, and after the delivery of the placenta, cuts the umbilical cord with her carnassial teeth (Houpt and Wolski, 1982). The cleaning is stimulated by the smell and taste of the amniotic fluid, and tends to be more effective in second and subsequent litters than for the first. Between individual births, the queen may be very restless, pacing around the nest. If a placenta is undelivered while this is occurring her movements will tend to stretch the umbilical cord, reducing blood flow through it so that when it is severed, it bleeds very little. Both placenta and umbilical are eaten; at this point a solitary queen faces several days with little or no food, so she may gain significant nutritional benefit from this. Inexperienced or incompetent mothers may occasionally appear not to distinguish the kitten from its placenta, and will eat the kitten as well (Deag *et al.*, 1988).

After the delivery of the last kitten the queen encircles her litter, usually lying on her side, and encourages them to suckle by nuzzling and licking them. Apart from occasional changes of position, she will remain in contact with the kittens for at least the first 24 hours. At this stage the mammary glands are producing colostrum, which is rich in antibodies, before it is replaced by the milk. For the first three to four weeks the queen spends up to 70% of her time in the nest, and takes complete care of the kittens. She initiates their feeding bouts by lying down beside them, and grooms them regularly, paying particular attention to their perineal areas to stimulate urination and defaecation. She either eats the wastes, or deposits them well away from the nest area. As the kittens become more mobile, they may stray out of the nest; the queen responds to the cries of a displaced kitten by grasping it by the scruff of its neck and carrying it back to the nest. This retrieving activity usually peaks in the third week. The queen is highly responsive to these cries, and will pick up and return with kittens that are not her own if they are in the vicinity of her nest. By this stage she will be spending more time away from the kittens, particularly if she has to forage for her own food, and will announce her return by the 'chirp' call.

From the third week onwards the kittens initiate an increasing proportion of the nursing bouts themselves, and soon after this the queen begins to bring dead prey items to the nest. Later, she will introduce them to prey that is still alive, putting them into a situation where they can refine their own prey-catching skills. She will also demonstrate the technique of burying faeces and urine, and the correct places to do so. In feral cats, the bringing of solid food starts the process of weaning, although human intervention often complicates the progress of this stage in litters born to family pets or show queens. From about day 30 onwards, the queen becomes less tolerant of the kittens, and may place herself out of their reach for extended periods of time. When with them, she may discourage them from suckling by adopting one of two postures, the crouch (ventrum in contact with the ground, all four paws supporting the body with the pads in full contact with the ground) and the lie (ventrum in contact with the ground and supporting the weight of the body, the legs partly extended or tucked under the body) (Deag *et al.*, 1988). Once the kittens are mobile they will attempt to ambush the mother away from the nest in order to suckle, and at this stage it often appears as if mother and offspring are engaged in a battle of wills. However, if the kittens are not removed the mother will often continue to demonstrate maternal behaviour towards them even when they are grown up, possibly more so if they are her daughters.

The antagonism between kittens and mother at the time of weaning is entirely understandable from a functional point of view. Provided the mother still has milk, the kittens will benefit from continuing to suckle, rather than test out their predatory abilities, and expose themselves to danger outside the nest. On the other hand lactation is highly costly to the queen, something that is easily forgotten when considering pet cats. The precise costs will depend on the condition of the queen at the time of parturition, and the size of the litter. If the queen has been malnourished during the period of gestation, she will tend to perform less active mothering of her kittens, and will be more aggressive towards them (Martin and Bateson, 1988). In litters greater than four, the birth weight of the kittens declines with number. The queen's milk production increases with litter size, but not in direct proportion, with the result that by eight weeks of age the kittens in a litter of six are about 25% lighter than those in a litter of

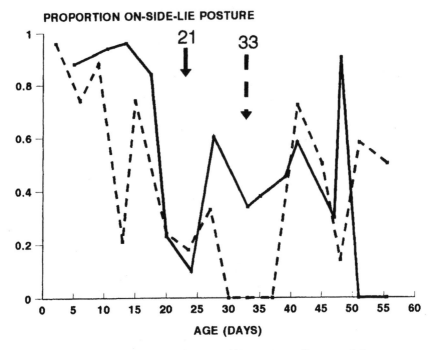

PROPORTION ON-SIDE-LIE POSTURE

Fig. 4.1. The relationship between the age of kittens in two litters, and the proportion of time that their mother adopted the nursing position. In both cases a discontinuity in weight gain occurred, on the day indicated by the arrows, which corresponds to a temporary decline in nursing. (Redrawn from Deag *et al.*, 1988.)

two. The mother can lose 6 g of body weight per day, or more with a large litter, and severe weight loss at this stage can also lead to increased aggression towards the kittens, although this may be reduced again if the queen's food is then supplemented. It appears that with larger litters the queen forces the weaning process, because there is a direct correlation between a 'dip' in the rate at which the kittens gain weight and a period of a few days in which the queen does not adopt the normal nursing position (Fig. 4.1). After this time, she will nurse them again, but usually the kittens will initiate this (Deag *et al.*, 1988). The chief factor seems to be the size of the litter, rather than the sex of the kittens; although males grow up to be heavier than females, there is no difference in their weights until after eight weeks, when the males begin to grow faster. In small litters there is often no discontinuity in weight gain, and in behavioural terms the mothers seem more tolerant of their kittens, the exception being the litter of one, when the mother is more aggressive towards her kitten after the normal weaning time than she would if there were two. Large litters may cause discomfort to the queen by pushing and biting at her nipples, exacerbated by the competition between littermates, making her less inclined to nurse. In functional terms, the queen will have a greater investment in each kitten if it is part of a small litter compared to a large one, and in the former case it may pay her not to risk inducing a metabolic stress that could coincide with exposure to an infection. Certainly kittens from smaller litters tend to go on suckling well beyond eight weeks of age, although it is not known how much nourishment they obtain from this. Suckling behaviour can persist into adulthood; instances have been reported of grandmothers nursing their daughters who themselves are nursing their own kittens.

Development of the Kitten

When it is born, the kitten is blind, virtually deaf, and completely dependent on its mother. From the time it is eight weeks old, it is capable of an independent existence, even though it may often stay with its mother for a longer period. Such rapid development could only occur if the maturation of the various sensory, motor and control systems were highly co-ordinated, but the system is not entirely rigid. External events can speed up or slow down individual processes; to take a physiological example, the development of body temperature regulation can be hastened by exposure to cold at two weeks of age, but by four weeks old all kittens have similar thermostatic abilities whether or not they have been exposed to cold. This adaptability suggests that each system has a goal of development, which it can achieve from a variety of starting points, by several routes, and by speeding up or slowing down different components, even to the extent that they end up occurring in different orders (Martin and Bateson, 1988). This flexibility is presumably built in to cope with the vagaries of weather, litter size, food availability, behaviour of the mother, and so on, so that a competent adult can emerge despite setbacks along the way. The following account therefore applies only to a 'typical' kitten, although ranges as well as averages will be given where useful, and the remarks made above about the bias towards single mothers and confined litters, made in connection with maternal behaviour, apply equally here.

The gestation period of the cat is about 63 days, and kittens in small litters weigh about 105 g when born. Litters may consist of between one and ten kittens, with a median of four or five, although in the wild the number is likely to be reduced by virus infections, to which the kittens are often highly susceptible. Compared to many other mammals, their development is slow, and involves a long period of maternal care. When newly born, the kittens crawl along the mother's abdomen to find a nipple, and alternate between suckling and lying still. Suckling can occupy up to eight hours per day at first, although this declines as the kitten matures. The kittens often purr while suckling, and tread or 'paddle' around the nipples, presumably to stimulate ejection of milk. If they are unable to locate the mother, or if they become isolated, cold or trapped (beneath the mother, for example), they use a 'cry' call to stimulate her attention; it is this cry, rather than their visual appearance, that identifies the kittens to their mother. The queen normally has eight nipples, of which those at the rear are often preferred by the kittens. Individual teat preferences appear quickly, and persist until weaning is under way, when the kittens will again suckle from any nipple. For the first three weeks, the mother initiates all the bouts of suckling, but from then on until the end of the first month the kittens themselves take an increasing role in this, until in the second month of life it is the kittens who initiate suckling, and the mother who begins to avoid their advances. They may even try to suckle while she is standing or moving about, and she may then try to block them, or even become aggressive. At the same stage the kittens become much more exploratory, and also begin to exhibit social behaviour towards each other, both play (discussed below), and other social patterns such as flank rubbing on the mother.

Social contact with the mother during the first four weeks is essential to the normal emotional development of the kittens, which otherwise develop a variety of behavioural, emotional and physical abnormalities. They are unusually fearful of other cats and people, exhibit randomly directed locomotory behaviour, and are slow to learn simple associations, such as the location of a source of food. Some of the stimulation provided by the mother can be replaced by simply handling the kitten for 20 minutes each day, which has an accelerating effect on eye opening, exploratory behaviour, and the development of mature EEG patterns. As has been mentioned above, maternal attention to the kittens can be reduced if the mother is malnourished. If her food intake is 80% of normal, the kittens are buffered from any direct nutritional effects and gain weight as normal, although they do spend more time nuzzling their mother's nipples. More severe deprivation of the mother results in kittens with smaller brains, particularly the cerebrum, cerebellum and brain-stem. At four months old, male kittens from these litters are more aggressive than usual, and both males and females have motor deficits that cause them to run erratically. The latter at least seem to be primarily affected by the shortage of protein in the diet, rather than the number of calories available (Martin and Bateson, 1988).

General activity does not increase smoothly until it reaches adult levels, but proceeds in a series of steps (Levine *et al.*, 1980). There is a sudden increase between nine and 14 days, reflecting the maturation of the auditory system, the opening of the eyes, and the beginning of walking. A second step occurs during the fourth week, when running is added to the locomotory repertoire, and a third towards the end of the second month, as locomotor play becomes more intense. The pattern of sleep also changes over the first two months. The total time awake stays roughly constant, at

30–40% of each 24 hours, but between three and six weeks of age the periods of wakefulness become longer. In very young kittens, the sleeping EEG is similar but not identical to that of paradoxical or dreaming sleep. The EEG indicative of quiet sleep gradually appears until, by the seventh or eighth week, the proportion of quiet to paradoxical sleep is the same as in the adult.

Reflexes

At birth, the kitten has had very little opportunity to learn, and so its behaviour largely consists of simple reflexes (Villablanca and Olmstead, 1979). Most of these are directed towards obtaining milk from the mother. The kitten locates the nipple and is stimulated to suckle by a variety of cues; the smell of the mother and her warmth guide them to the ventrum, tactile stimuli detected by their lips guide them on to the nipple, and tactile and chemosensory cues on the tongue stimulate suckling. Nipple attachment is achieved by the rooting reflex, in which the head is pulled back and then lunged forward while the mouth is opened. Once the nipple is located the suckling reflex, which is already present 12 days before normal parturition, takes over.

Neonatal kittens are capable of a certain degree of locomotion, which they achieve by pulling themselves along with their front legs while paddling with their weaker hind legs. During such movement, which may be random or directed towards sources of warmth, the head is often turned from side to side, presumably in an attempt to locate the mother. By the fourth day of life kittens are capable of travelling up to half a metre on their own, orientating themselves by olfactory cues derived from their own body smells. This ability is put to use if they are displaced from their nest because they have continued to cling on to their mother for a few seconds after she has got up and left. The motor cortex for co-ordinated forelimb movement matures during the first two weeks, and for the hindlimbs during the next two, so that rudimentary walking begins at about three weeks of age. At the same stage, the reflexes for ground sniffing and the crawling are disappearing. Adult patterns of locomotion do not appear until the seventh week.

One reflex that can be readily seen is flexor dominance of the vertebral musculature, which is triggered when the kitten is picked up by the 'scruff', the loose skin on the base of the neck. The limbs become limp, the tail is curled, and on release the kitten appears to be startled by its new surroundings, as if its sensory systems had been suppressed while it was being carried. It is presumably adaptive for a mother carrying a kitten to assess and react to any danger that occurs, rather than have the kitten react also and risk being dropped. This reflex persists into adolescence and can still be triggered in many adult cats.

Voluntary urination and defaecation appear at about three weeks of age,

and have completely taken over from the mother-stimulated reflex elimin-
ations by seven weeks. The air-righting reflex appears at about 24 days
(range 21–30), and is perfected by about 40 days (range 33–48). The
orienting reflex is first seen between eight and 12 days, and begins to be
associated with investigatory behaviour from one to four days later.

The 'gape' response to cat urine appears at about five weeks of age, and
is fully expressed at seven weeks. Since kittens have well-developed olfactory abilities considerably before this, it is possible that the gape depends
on maturation of the vomeronasal organ. Another stereotyped response to
a species-specific signal, piloerection towards the silhouette of a threatening cat (tail down, arched back, erect ears), is complete by six to eight
weeks (Kolb and Nonneman, 1975), although certain components,
including the arched back, are present at birth, and piloerection and pupil
enlargement appear at about three weeks.

The senses

Kittens are born with only four senses functioning efficiently (Fig. 4.2).
Tactile sensitivity appears about half way through gestation, and the vestibular balance organ functions even in kittens born a week prematurely.
Olfaction is present at birth, and is immediately essential for guiding
suckling behaviour. By three weeks of age the sense of smell is essentially
mature. Taste may be less important, because kittens less than three weeks
old will suckle from non-lactating females. Hearing, and then vision,
develop after birth.

Hearing

At birth, the auditory system is capable of transducing sounds, but the
canal is blocked by ridges of skin. These gradually open up as the canal
itself widens, over the first 10–15 days. Over the same period the pinna
enlarges and becomes corrugated at its base. The pinna can be moved
almost from birth, initially both spontaneously and also in a diffuse way to
a wide range of non-auditory stimuli including tactile, visual and olfactory.
By the third or fourth weeks the pinna movements become more discrete
and adult-like. The first responses to sounds come on about day five,
before the auditory canal is open, and consist of general arousal (sniffing,
lifting the head and head movements), bouts of pinna movements, and
squinting of the eyes. Species-specific calls, such as the kitten call of the
nursing mother, the growl, or the kitten distress call, elicit responses more
effectively and earlier in development than artificial sounds do. By day 16
directed head movements make it clear that the kitten can locate the source
of a sound. By three to four weeks old the kitten shows that it can discriminate between different kinds of calls; it tends to approach kitten and

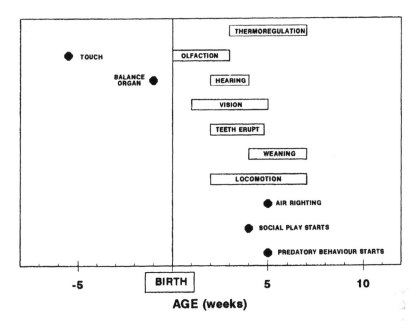

Fig. 4.2. Approximate timings for the appearance and development of some sensory and behavioural abilities in kittens. (The development of taste probably parallels that of olfaction.) (Data from Martin and Bateson, 1988.)

mother calls, but withdraws with ears flattened from growls and male threat calls (Olmstead and Villablanca, 1980).

Vision

The kitten's eyes open at between two and 16 days after birth. The chief source of this variability seems to be genetic, although the age of the mother and the sex of the kitten also have an effect; female kittens open their eyes earlier than males. At the moment of eye opening vision is barely functional, partly because the optic fluids do not become completely clear until another four weeks have elapsed, and partly because the visual cortex requires visual experience before it can mature. Therefore, the acuity of the eyes is initially poor, but improves 16-fold between the second and tenth weeks. Visual cues gradually become important over this period; kittens can usually recognize their mother by her appearance, rather than by her smell and her warmth, by the end of their third week (Martin and Bateson, 1988).

It may seem surprising that a sense as important as vision does not mature spontaneously in the cat. The requirement for visual experience seems to be driven by the cat's binocular vision, because in the rabbit, and other mammals that have almost panoramic vision, the visual abilities are only slightly affected by early experience (Blakemore and Van Sluyters, 1975). A certain amount of flexibility seems to be necessary to allow matching of the receptive fields in the two eyes, so that the neural mechanisms for stereoscopic vision and depth perception can be built up. At eye opening, many neurones in the central cortex will respond when either eye is stimulated. If the visual signals from the two eyes are widely disparate – for example, the kitten is born with a severe squint – these cells lose their binocular properties, and segregate into two groups, approximately equal in number, each of which responds to only one eye. The exact matching of the images from the two eyes is achieved by the emergence of the orientation-selective neurones (see Chapter 2) which respond to high-contrast edges. In kittens under three weeks old the orientation-selective cells are very broad banded by adult standards, and therefore presumably provide rather poor binocular vision. However, it seems that this initial approximation is enough to stimulate the development of much more specific cells, which appear very rapidly and are essentially adult in character by the time the kitten is four or five weeks old. Some of these cells respond to spots of contrast, others to vertical contours, and yet others to horizontal contours. If any of these elements is completely missing from the kittens' environment (which could only be achieved by deliberate deprivation) the corresponding neurones fail to develop, and it has also been shown that the cat must not only see these features, but react to them behaviourally for the complete set of orientation detectors to develop. The infantile neurones, now redundant, degenerate, and once the kitten is three months old the adult pattern cannot be reversed, nor can it be induced if it has not already appeared. This period from eye opening to three months is known as the 'critical period' for vision in the cat. At one time it was thought that the development of many behavioural patterns could only take place in critical periods, but these have proved difficult to demark, particularly for mammals, and have been replaced with concepts of goal-directed development, as mentioned at the beginning of this section.

The initial co-ordination of vision with movement is largely achieved in the second half of the first month of life (Norton, 1974). Kittens begin to orientate towards visual stimuli about five days after they are able to orientate towards sounds (Table 4.1). It is possible that the latter ability only appears when motor development allows, but if so, the former must therefore result from a maturation of the visual sense, or its connection through the visual cortex to the motor cortex. Recognition of the three-dimensional nature of the visual world develops later still, at about four weeks of age. Depth perception is often tested using a visual cliff. In this simple apparatus, a sheet of glass is placed over the edge of a step, usually about half a metre high; the horizontal surfaces, both before and after the vertical drop, are painted with a checkerboard pattern. As the kitten

Table 4.1. Ages at which various orientated reactions occur to some auditory and visual stimuli. Descriptions of the visual cliff and visual placing tests are given in the text.

Response	Age of kitten in days	
	Earliest positive response	Latest negative response
Orientating to a sound	11	16
Following a moving sound	15	21
Orientating to a visual stimulus	16	21
Following a moving visual stimulus	18	24
Avoiding a visual cliff	25	37
Reflex visual placing of forelimbs	27	36
Avoiding obstacles around the nest	26	35
Guided visual placing of forelimbs	33	39

Adapted from Norton, 1974.

approaches the edge of the step, it should be able to see the drop in front of it, but at 24 days old it will almost always continue its progress, supported by the glass surface. Sometime over the next two weeks it will begin to hesitate on the 'edge', and turn back. At almost exactly the same age the kitten will also learn to take avoiding action as it blunders about the area of its nest, so it seems that around this age the connection is established between images and the objects that produce them.

Also at about this time, co-ordination between eyes and limbs develops (Table 4.1). This can be simply shown by the visual placing response. If a kitten is lowered gently towards a visible surface, it will extend its forelimbs automatically, as it would when landing from a jump. Initially the response is undirected; the kitten sees the surface, and extends its legs down in a reflex manner; if there are holes in the surface, it is just as likely that a paw will go through one of these as land on a solid area. Not for another week or so are the paws actually guided on to the surface by direct co-ordination between eye and limb (Hein and Held, 1967).

Learning abilities of the kitten

The way that the development of visual abilities depends on actual visual experience is in itself a kind of learning. From the very first days, kittens show that they are also capable of making associations between stimuli in their environment. The first overt indication of this is usually the appearance of nipple preferences, guided by olfactory and textural cues from the mother's ventrum. Using an artificial mother, consisting of a carpeted surface with two rubber nipples, Rosenblatt (1972) has shown that a two-day-old kitten can learn to distinguish between a nipple that delivers milk, and one that does not, based on its texture alone. Discrimination based on odour is possible just one day later. Suckling improves with practice, both in terms of how effective it is in obtaining milk, and how well the kitten discriminates between its own mother and other sources of milk (artificial nipples were tested, but this suggests that a kitten can also discriminate between its own mother and any other lactating females if it is in a pooled litter).

Learning also plays a crucial role in orientation back to the nest area. For the first few days, an isolated kitten will crawl in circles wherever it is, but from about day six this only occurs in the nest; elsewhere, despite still being blind, they can orientate themselves towards the nest using learned olfactory cues, although their initial strategy is usually to cry to attract their mother's attention. Odours from the mother or from littermates are both adequate signals. At two weeks old, the same mechanism is still used, but is effective over longer distances, of up to 3 m. From then on the kittens will leave the nest voluntarily for short periods, but it is not until they are three or four weeks old that visual cues dominate over olfactory ones in guiding them home.

Learning by observation

The mother's influence is not restricted to the provision of food and protection, for in the extended period that the litter stays with her she does a great deal to stimulate their learning. We have already seen that adult cats can learn tasks more quickly if they can observe another, more experienced, adult performing that task, so it would be surprising if kittens did not also learn by observation. It has indeed been shown that kittens can learn to perform an arbitrary task (pressing a lever in response to a flickering light) very quickly if they can watch an adult perform the same task, while kittens that are presented with the task without ever seeing an adult perform it learn very slowly, if at all (Chesler, 1969). Such a large difference in speed of learning might not have been apparent if the task had been based on more ecologically relevant motor patterns, but it does appear that the opportunity to observe adults does promote learning in kittens. In the same series of trials, it was found that mothers were more effective demonstrators than other females. This was particularly marked in terms of the number of sessions taken before the kittens began to succeed (Fig. 4.3); once they began to perform correct responses, the time taken to improve to giving the correct response on every occasion was the same whether their mother or another cat was the demonstrator. The difference appeared to lie in how much attention the kitten initially paid to the demonstrator. Since it

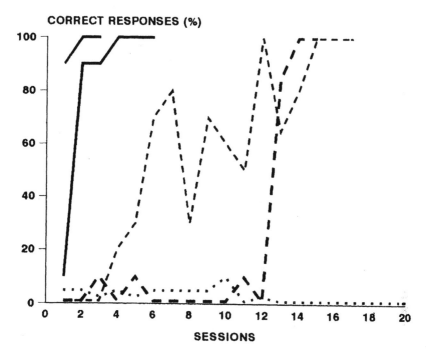

Fig. 4.3. The acquisition of a response (pressing a lever) in representative kittens from three groups given different opportunities to learn. The solid lines indicate the proportion of correct responses in kittens allowed to observe their mothers completing the same task; the broken lines, kittens that observed a stranger; the dotted line, a kitten that was given no opportunity to observe. (Redrawn from Chesler, 1969.)

might be expected that a kitten would initially be apprehensive in the proximity of a stranger, this explains the greater attention paid to the mother in the first few sessions.

The development of predatory behaviour

The most important skills that the mother demonstrates to her kittens are those they will use later for hunting. From the beginning of weaning onwards, she brings recently killed prey to the nest; later, the prey is brought in alive and released near to the kittens. In this way she puts them in a position in which their innate responses, many similar in form to those used in play, bring them into contact with prey at an age at which they would find it impossible to achieve the same degree of contact on their own. Thus the goal of hunting, and the link between predatory behaviour and food, are firmly established at an early age. The role of the mother in this is part demonstrator, and part supervisor, in that she sets up the situation in which

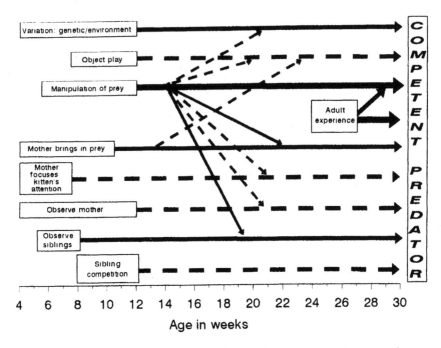

Fig. 4.4. Some of the factors that are thought to influence the development of predatory skills in adults. Proven influences are shown by solid arrows, and hypothetical influences by broken arrows. Diagonal arrows indicate some of the possible interactions between influences. The approximate duration of each influence is indicated by the width of its box. (Redrawn from Martin and Bateson, 1988.)

the kittens can learn, but does not necessarily perform the required patterns in front of them. However, if the kittens hold back from attacking the prey, she is likely to initiate an attack herself (Caro, 1980).

The kittens' tendency to follow their mother's lead extends to food choice as well. They imitate their mother's food selections, so that if the mother has previously been trained to eat an unusual food, such as banana or mashed potato, they will also prefer this to more 'natural' foods. The peak of imitative learning of food choice comes when the kittens are seven or eight weeks old, and after this their preference for whatever their mother ate persists even when she is not present. A similar process probably takes place with natural prey, because kittens prefer to kill strains of rat that they have seen their mother kill, over other strains. This may be an effect largely confined to large prey with the capacity to fight back and cause injury to a young, inexperienced cat, for Adamec *et al.* (1980) found that observation of prey killing had much less effect on subsequent skill if the prey was a mouse than if it was a rat. Indeed, a single experience of killing and catching a mouse can turn a kitten into an accomplished mouse-killer. The main effect of observation in the case of large prey may be to overcome the fear that conflicts with the predatory drive in inexperienced cats, and causes them to react defensively towards rats.

The actual behaviour patterns that kittens use to interact with prey have been classified as 'predatory play', but since such sessions would, in the wild, almost inevitably end with the consumption of the prey, this may be stretching the definition of play (see next section). However, virtually all of the motor patterns seen in object play also form part of prey handling, including Poke/Bat, Scoop, Grasp, Bite/Mouth, Belly-up, Stand-up and Pounce, so the link between the two sorts of activity is evidently a strong one. Skill in prey handling during the weaning period (one to three months) accurately predicts predatory abilities at six months, and early weaning accelerates the onset of actual killing of live prey. However, the origins of the skills themselves are less easy to pinpoint. This is probably because a multiplicity of influences affect the development of hunting ability (Fig. 4.4), and indeed one would expect that, for such an essential function as obtaining food, development would need to be highly flexible to cope with the varied ecological situations into which different kittens may emerge.

Play Behaviour

Play is easily recognized, but difficult to define (Martin, 1984). Some of the behaviour of young animals brings about immediate consequences that can reasonably be called adaptive, suckling is an obvious example. It is not so easy to find an immediate benefit for many other activities, such as manipulating inanimate objects, chasing littermates, or climbing on to obstacles around the nest (although attempts have been made to do so), and such activities are often placed in the category of play. One formal definition is 'all motor activity performed postnatally that appears to be purposeless, in which motor patterns from other contexts may often be used in modified forms and altered temporal sequencing'. Obtaining hard information on the benefits that playing as a kitten brings to the adult cat has proved difficult, and so for practical purposes play is often defined in terms of the motor patterns involved, which are often similar to species-specific behaviour patterns performed by adults, although the context, intensity and sequencing are all likely to be altered when they form part of play.

Perhaps the most convenient way to classify play behaviours is in terms of the object towards which the kitten directs such motor patterns. Social play involves a conspecific, which is likely to be a littermate but can also be the mother or other adult. Play can also be focused on inanimate objects; these are treated differently depending on whether they are large and immobile, in which case they are climbed on and jumped off, or small and mobile, in which case they appear to become substitutes for prey items. The latter type of object play can also be seen directed at invisible objects; for example, kittens may pounce on 'thin air'. It is impossible to tell in such cases whether the kitten really has imagined a prey item, or whether a speck of dust has triggered a set of behaviour patterns normally reserved for larger objects. Kittens also chase their own tails, such self-directed play forming a category that has received little attention from cat behaviourists. However, locomotor play, object play and social play have all been studied in detail in the cat, and each will be described before their possible functions are discussed.

Locomotor play

Adult cats are extremely agile, and those parts of the nervous system concerned with balance and locomotion are correspondingly well developed. Some of the skills that contribute to agility appear almost fully formed, for example the air-righting reflex, whereas others develop more slowly, and seem to improve with practice. Once they have left the nest, kittens explore their environment vigorously, not only by looking, sniffing and listening, but also by clambering over any objects in their path, including those that are unstable or will not bear their weight; gradually they build up experience of which types of objects are safe to walk or climb on, and which are not. Much of this exploratory locomotion appears to have little short-term benefit (it may even put the kittens at risk by making them conspicuous to predators) and so may be classified as play; certainly it appears to be rewarding to the kittens for its own sake.

This aspect of play has not received the same attention as object play and social play, but is nevertheless a crucial aspect of the kittens' development. A study by Martin and Bateson (1985) described many of the important features of locomotor play. Seven litters of kittens were given access to a wooden climbing frame every few days, from the time they were 36 days old, to the end of their second month. To begin with they spent only about three minutes of each half-hour test period on the frame, but by seven weeks old this had increased to almost 20 minutes. For comparison, their mothers spent between one and five minutes on the frame during the same periods. The frame was used for social as well as locomotor play, but the latter predominated, possibly because opportunities for social play were also available outside the test sessions. Until they were about seven weeks old, the kittens largely restricted their activities to the lower rungs of the frame, presumably because their locomotor systems were not sufficiently mature until then. Some had still not ventured to the highest parts by the end of their second month, and those that did tended to lose their balance more often than those that did not, although such slips rarely resulted in any kitten actually falling off the frame. After these occasional falls, the kitten concerned would normally climb straight back on to the frame, confirming the self-rewarding nature of such play. Male kittens were no more or less exploratory than females, but there were considerable differences between families. Some of these differences appeared to be linked to the mothers' behaviour; kittens which eventually made use of the top of the frame were those with mothers which had spent the most time on the frame during the first few sessions.

Object play

Confronted with novel objects, kittens will investigate them carefully by looking, sniffing, licking and touching, and will often circle around the object to repeat this process from a number of angles. Subsequently, small objects are usually played with by being poked, batted, grasped and tossed in the air. Such bouts of play are often preceded by pounces or leaps reminiscent of the predatory behaviour of adults. Larger objects can be used as hiding places or perches, and for locomotor play.

Descriptions of some common object play patterns are included in Table 4.2. Most of these are not performed in isolation, but form sequences in which the various patterns occur in predictable, though not fixed, orders. Bouts of object play tend to go on for much longer in felids than they do in the young of other carnivores, usually because similar sequences are repeated over and over again. The Poke/Bat pattern may be crucial to stimulating this repetition, because it makes the object move, whereupon it can be chased and pounced on (Aldis, 1975). In one study, 35% of play bouts were started by Poke/Bat, which also made up 29% of patterns observed within sequences (West, 1979). Poke/Bat was also often

Fig. 4.5. Three postures seen in play; (left to right) Bat, Grasp, Side-step (see Table 4.2).

Table 4.2. Some definitions of behaviour patterns seen in kitten play. Not all authors have used precisely these categories; for example, similar, but not identical, descriptions can be found in Barrett and Bateson (1978) and Caro (1981). The letters O and S denote patterns that form part of object and social play respectively. The numbers in brackets indicate the time that each social pattern first appears (West, 1974).

Scoop	O	Kitten picks up an object with one of its front paws, by curving the paw under the object and grasping it with its claws
Toss	O	Kitten releases an object from the mouth or paw with a sideways shake of the head or paw
Grasp	O	Kitten holds an object between the front paws or in the mouth (Fig. 4.5)
Poke/Bat	O	Kitten contacts an object with either of the front paws from either a vertical (poke) or a horizontal (bat) orientation
Bite/Mouth	O	Kitten places an object in the mouth and closes and opens its mouth around the object (Fig. 4.5)
Belly-up	O, S	Kitten lies on its back, belly up, with all four limbs held in a semivertical position; the hindlimbs may move in treading movements and the forelimbs may be used to paw at an object or kitten; the mouth is typically open (Fig. 4.6) (22 days)
Stand-up	O,S	Kitten stands near or over an object, or another kitten (the target); if the latter, its head is oriented towards the head and neck region of the other kitten; the kitten's mouth is open and it may lunge with its mouth, or raise a paw, towards the target (Fig. 4.6) (24 days)
Vertical Stance	O, S	Kitten extends its back legs so it is in a bipedal position with the forelimbs outstretched (36 days)
Pounce	O, S	Kitten crouches with its head touching the ground or held low, its back legs tucked in and tail straight back, often moved back and forth; kitten then treads with its hindquarters before leaping forwards and upwards by rapidly extending its hind legs (34 days)
Chase	O, S	Kitten runs after a moving object or kitten (39 days)
Side-step	O, S	Kitten arches its back, curls its tail upwards and towards its body, and walks sideways (Fig. 4.5) (33 days)
Horizontal Leap	S	Kitten arches its back, curls its tail upwards and towards its body, and leaps off the ground (43 days)
Face-off	S	Kitten sits near another kitten and hunches its body forward, moving its tail back and forth, and lifts a paw towards the other, often with a swinging motion (45 days)

From West, 1979.

followed by another Poke/Bat, with as many as 11 repetitions being recorded. The other common initiating pattern was Vertical Stance (21%), which also made up 15% of behaviours within sequences. Other common components were Bite/Mouth (12%), Chase (10%) and Belly-up (9%). The Scoop pattern is less common than any of these, but has an interesting variation, seemingly preferred by cats but by few other carnivores, in which it is performed through a hole at an object on the other side. In general, although object play has been compared to predatory behaviour, the sequences are more reminiscent of the adult cat's post-hunting manipulation of prey, which is often thought of as a displacement activity, and will be discussed further in Chapter 7.

Social play

Kittens begin to initiate play with their littermates as soon as their sensory and motor systems are mature enough, and continue to interact at a high rate for the next few weeks. Many of the behaviour patterns seen are similar to those used in object play (Table 4.2), modified to take account of the liveliness of the target, while others are only seen as components of social play. The latter, for example the arched-back component of Side-Step and Horizontal Leap, are similar to patterns of social interaction seen in older cats, whereas the former are more predatory in form. West (1979) described the most common patterns for initiating social play as Pounce (39%), Belly-up (14%) and Stand-up (16%). These were generally about 90% effective in eliciting a response from the other kitten, and were used in the same proportions at both six and 12 weeks of age. Two patterns that changed in frequency of use were Vertical Stance (8% at six weeks, 24% at 12 weeks) and Side-step, which declined from 20% to 3% over the same period. Side-step was generally only about 75% effective at obtaining a response, so it is possible that between weeks 6 and 12 the kittens learned

Fig. 4.6. Two postures seen in social play; Variation of stand-up (left) and Belly-up (right). (M.T.)

to avoid using this behaviour pattern at the beginning of sequences, because other initiations were more likely to achieve the desired bout of play. Another unusual attribute of Side-step was that it tended to elicit a Side-step in the responding kitten, whereupon the initiator kitten would perform another Side-step, and so on. Other common patterns tended to elicit a complementary, rather than identical, response. So for example Pounce was often reciprocated by Belly-up, Belly-up by Stand-up, Stand-up by Belly-up, and Vertical Stance by Belly-up. Such alternations tend to bring about a reversal of roles within the pair of kittens as the interaction proceeds, each taking it in turns to be initiator and recipient. Bouts were most often terminated by two specific patterns, Chase and Vertical Leap.

The goal of such social play appears to be physical contact between the kittens. Since several of the typical behaviour patterns are similar to those used by adult cats during fights, it is worth considering why bouts of social play do not escalate into truly agonistic encounters. One simple answer is that they occasionally do, and this becomes more common as the kittens get older, and as elements of sexual behaviour become incorporated into the repertoire of male kittens. However, the majority of bouts are apparently amicable, and the observation that each kitten alternates between being the 'aggressor' and the target indicates that there is little, if any, true aggression involved. It has been suggested that certain signals indicate the intention to play rather than to fight. One is/ the half-open mouth, the kitten's 'play-face', often displayed at the beginning of the interaction, and during such patterns as Belly-up. Another is the ritualized nature of many of the patterns, always performed at the same level of intensity and in a stereotyped way, which should tend to enhance the confidence of the recipient. Signals produced by moving the tail at particular rates have also been suggested as indicators of an intention to play. Furthermore, several behaviour patterns that signify aggression in adults are not seen in the context of play, one example being the defensive arched-back posture, which is quite distinct from the arched-back postures (Side-step and Horizontal Leap) used in play.

All of the above presupposes that there are at least two kittens in the litter; occasionally only one is born, or disease leaves a single survivor. Social play in one-kitten litters has been studied by Mendl (1988); one interesting finding was that single kittens did not engage in any more object play or self play than did kittens in litters of two. This indicates that, despite the overlap in motor patterns between object and social play, they are separately motivated, a conclusion also supported by the ages at which each type of play is most intense (see below). The solitary kitten must direct all its social play towards its mother, and since the mother is much more likely to be absent from the nest than is a sibling, it is not surprising that the single kittens were involved in relatively little social play. However, they are able to compensate partly for what appears to be the preferred type of social play, with siblings, by initiating bouts of play with the mother, and the mother does respond with play behaviour patterns herself. This continues up to the time when the kitten is about two months old, when she begins to respond aggressively towards invitations to play. There is no evidence that the mother somehow 'realizes' that she should play with a single kitten because otherwise it will receive none; it is the kitten, not the mother, that initiates each play bout, and the mother may simply be responding to signals that would be directed almost exclusively at siblings in a larger litter. Also, the types of behaviour that the solitary kitten pursues are altered; it can usually stalk its mother, and paw her tail, without disturbing her unduly, but she seems unwilling to reciprocate the more intense patterns such as wrestling and chasing. If, as has been suggested, activities involving body contact are the goal of social play, then the quality of play for the single kitten may reduced. The long-term consequences of this for the kitten, in particular its social behaviour as an adult, are unknown.

The development of play

As has already been indicated for the behaviour patterns that initiate bouts of social play, the various components of play are not expressed in the same proportions as the kitten grows. Moreover, social play and object play, while overlapping considerably, peak at different stages of development. The precise timing of all these changes can be affected both by genetic factors, the behaviour of each individual mother, and environmental variables such as the availability of food. The typical timings given below should therefore be taken as approximate, and as only applying directly to kittens housed indoors with their mothers.

Many components of social play are first seen as isolated, self-directed patterns, often repeated over and over again as if being practised – the approximate times at which each appears are indicated in Table 4.2. In the fifth or sixth week mixed sequences appear, and are directed towards other kittens for the first time. Complex sequences develop rapidly, but the Vertical Stance and Face-off patterns are not common components of sequences until several more weeks have elapsed. During the first three weeks or so of social play, three or more kittens are often involved in the same sequence, but by about the ninth week the great majority only involve a pair. The peak of social play is a broad one, from about the ninth week to the 14th.

Object play occupies a smaller proportion of the kittens' time (although it may occur quite frequently, but briefly, during bouts of social play) until about week seven, when it increases in frequency and can become commoner than social play by about week 16, after which its frequency seems to depend considerably upon the 'character' of the kitten and the conditions under which it finds itself. This changeover occurs in parallel to changes in other types of social behaviour. For example, kittens approach one another more, touch noses more, and are more likely to be in contact with one another in weeks 10–14 than in weeks 18–21, but in the latter period they are more likely to be genuinely aggressive towards one another (as distinct from the 'mock' aggression of social play) (Mendoza and Ramirez, 1987). There appears to be a major change in the organization of play activity in the fourth month, which correlates with, but is not necessarily driven by, changes in the environment in which the kittens find themselves. From birth until the fourth month, the kittens are confined to a small area, because at least initially their motor capabilities are insufficient to take them far from the nest, and in any case they need to be close to their mother for nourishment. Siblings are therefore readily available, and opportunities for social play are all around. The fourth month is the time that feral cat mothers often leave their kittens, and coincident with this there is an increased tendency to explore the environment, and play in, on and with objects is evident, even in kittens that are still housed with their mothers.

There also appears to be a more subtle change in the quality of play at the end of the second month, about the time that weaning is normally completed. Comparing different litters, Caro (1981) found that some patterns, such as Pounce, Poke/Bat and Bite/Mouth (using the terms in Table 4.2), could be grouped together, and showed a positive association with the components of the emerging predatory behaviour, while others, including the arched back of Horizontal Leap and Side-step, Vertical Stance, and Chase, became progressively dissociated from predatory behaviour patterns. This may indicate that, while a single motivational system may control social play when it first appears, as the kitten develops certain components may be elicited differentially by two systems, one connected with predatory behaviour, and one with agonistic social behaviour. There are, of course, other explanations for some of these changes; a possible reason for the decline in the frequency of Vertical Stance, based on learning, has already been mentioned.

Several studies have examined variations in the process of weaning on the expression of play. In particular, the switch from social to object play seems to follow the completion of weaning, and might therefore be driven by this event. If the mother cannot obtain as much food as her normal *ad lib* intake, she will tend to wean her kittens earlier than usual, even if their growth is unaffected. This stimulates both object and social play in the kittens. The mother appears to achieve this directly, and at a much earlier stage than object play itself appears, by reducing some aspects of maternal care during the first three weeks after the birth (Bateson *et al.*, 1990). While one might predict that inadequate nutrition should lead to a reduction in a 'useless' activity such as play, in fact play is not energetically expensive (see below), and increased object play may be a way for the kittens to gain predatory skills more rapidly if it seems likely that food is going to be hard to find.

Sex differences in play behaviour

Until about 19 weeks of age there appear to be few qualitative differences in social play between male and female kittens. At that time the sexual behaviour of males begins to appear, and they will sniff the females' ano-genital areas, attempt the neck grasp and try to mount. Females often respond aggressively to such advances until they begin to reach sexual maturity, from about 23 weeks. Sex differences in object play emerge much earlier, from about week seven, when the rate of object play by male kittens begins to rise to about twice the rate performed by females (Barrett and Bateson, 1978). This is not due to a simple effect of motor development, since at this stage females are actually more active than males. The increase in object play is particularly marked in single-sex litters, since in mixed litters the females tend to play almost at the same intensity as males. A similar 'masculinization' effect is also detectable in social play between 12 and 16 weeks, such that all-female litters play slightly less aggressively than do mixed or all-male litters.

The functions of play

Although there has been a great deal of speculation about the functions of play in cats, and in many other mammals, little hard evidence has been produced to support any theory. The underlying notion has been that the play of the young animal somehow produces benefits later in life, although this has been challenged by some authors, who have speculated on benefits which the young animal might gain almost immediately, and which might be undetectable in the adult. The search for adaptive benefits has also tended to assume that play involves the young animal in considerable costs, which must be recouped later in life, otherwise kittens that did not play would tend to survive to adulthood, breed, and perpetuate whatever genes were available to suppress play. In fact, it has been shown that the metabolic costs of playing are slight; a kitten that plays for a typical 9% of the day will increase its energy expenditure by an average 4%. There may be other costs to free-living cats, such as increased exposure to predators during locomotor play, but these have not been quantified.

The presumed functions of play can be divided into three categories: motor training, cognitive training and socialization. Much of the research on cats has focused on the former, and in particular on the impact of play on subsequent predatory skills. Comparison of the cat with the dog shows that many differences in their play behaviour can be directly correlated with differences in their predatory behaviour. For example, puppies indulge in violent bouts of shaking of objects held in their mouths, reminiscent of the killing shake of wolves, whereas cats rarely do so. Object play and hunting are both solitary activities for cats, but for dogs these are social and co-operative activities respectively. Stalking and ambushing are important parts of both social and object play in cats, but dogs rarely use these in hunting or perform them during play (Aldis, 1975). There are at least two possible interpretations of these similarities, however. One is that performance of predatory motor patterns during play improves the motor skills of the young animal, so that it becomes a more efficient predator later in life, and it is therefore logical that practice of the predatory motor patterns themselves is the best preparation for the real thing. Another explanation for the similarity between play and predation is simply that the predatory motor patterns are already to some extent programmed into the motor, sensory and control systems of the young animal, and it would be inefficient for a complete set of play-specific patterns to be programmed as well, particularly as they would, by definition, only be active for a few weeks in the whole lifetime of the animal. Thus, if possible, play should 'borrow' motor patterns from those used universally by adults. Following this line of reasoning further, even if the primary function of play were not the practice of predatory skills, we might expect to see predatory patterns used, because they are already in the cat's repertoire, and with some modification could be used to facilitate cat–cat interactions.

Certainly it has been very difficult to prove any major link between the degree of object play or social play in kittens, and subsequent predatory skills in those same individuals. This is probably because there are all kinds of factors that could, or have been shown to, influence the competence of adult hunting, and none is in itself essential (Fig. 4.4). Since kittens in the wild are at great risk immediately after weaning, it is possible that it is only at this stage that play has a significant effect on hunting success; by the time the kittens that have survived reach adulthood, those that were less competent at weaning have caught up, and so no effect of play is then detectable (Martin, 1984). This fits in with the apparently goal-directed nature of much kitten development, discussed earlier, and with the observation that early weaning stimulates object play.

Less attention has been paid to the possible role of play in forging cognitive skills that will be used later as components of social behaviour. This is unsurprising in view of the comparatively recent discovery of the complexities of adult cat sociality. However, the separation of social play into aggressive and predatory components, probably under the control of separate motivational systems, suggests in itself that this type of play may have functions other than the honing of predatory skills. Establishing links between social play, and social behaviour late in life, will make a fascinating area of research for the future.

Finally, West (1979) considers that we may be looking too far into the future to discover one function of social play; it may simply serve to keep the litter together at a time when each kitten needs to be easily found by its mother for suckling, and when an isolated kitten may present an easy target to a predator – play as an 'invisible play-pen'.

Changes in Old Age

Adult life is not a behavioural plateau that follows the steep upward slope of kittenhood. Changes occur in the central nervous system as the cat ages that are almost certainly reflected in behaviour, although these have hardly been studied at all compared to those of the young animal. Conditioned reflexes (see Chapter 3) are harder to establish in older cats than in younger, but cats between 11 and 16 years old are actually more mentally flexible than younger animals, for example in learning changes in the location of food. If they have remained healthy, their locomotory skills are unimpaired at this age, as are their reactions to the vocalizations of other cats (Levine *et al.*, 1987). However, they habituate much less quickly to repeated stimuli than younger cats do. It is known that certain areas of the brain are susceptible to ageing, while others are much more resistant, and so selective changes in ability might be expected as cats age. It seems likely that a cat cannot be said to be functionally old until at least its 16th year.

Communication

Until quite recently, the communicative skills of domestic cats have been undervalued, because it was thought that the average adult cat only needed to communicate the boundaries of its territory, and its willingness or unwillingness to participate in aggressive or sexual encounters. The discovery of a rich and varied repertoire of sociality has presented new challenges to those interested in the mechanics of social interaction between cats, and a great deal still remains to be discovered in this area (see Chapter 8). The present systems of communication between domestic cats have arisen as a product of evolution from ancestral Carnivora, and the requirements of the habitat in which *F. s. lybica* developed. There may have been further changes as a result of domestication, but until the wild species is investigated in more detail it is impossible to determine what these might be. Certainly the stability of social groups of farm cats argues in favour of the retention of a full repertoire of ancestral communication patterns.

Particular modes of communication suit different sets of circumstances; before each is discussed in detail it is worth outlining the particular advantages and disadvantages of communicating by sight, sound or smell. Olfactory communication is especially important in solitary species that do not know when the message they wish to convey will be received. Odours can be active for long periods of time if they have been appropriately formulated, and are placed where they will not be destroyed by the effects of weather. Such signals cannot, however, be turned off at will, which may not present many problems for a large animal with few enemies, but may be a distinct disadvantage for a prey species. Orientation to chemical signals is at best chancy, because over distances of more than a few centimetres they are carried by the wind, and when solitary cats need to locate one another, for example when a female is in oestrus, sounds are used as well as odours.

Table 5.1. Properties of the three main signalling channels when used over long distances.

	Maximum range	Direction	Sensitivity (signal/noise)	Source location	Signal off
Sight	Medium	Straight	Medium	Direct	Rapid
Sound	Long	Omnidirectional	High	Indirect	Rapid
Odour	Long	Of wind	Very high	Very indirect	Slow

Chemical signals are most effective when used at night, or underground, or in dense vegetation, circumstances in which visual communication is difficult. Auditory signals are as effective at night as during the day, and depending on their frequency can be made easy or difficult to locate. They are also effective over very long distances, one example from another carnivore being the wolf's howl. Over short distances during daylight almost any medium can be effective, and when unambiguous communication is critical to survival, such as during agonistic encounters, several signals are used simultaneously. Over long distances each modality has distinct properties, some of which are listed in Table 5.1.

Acoustic Communication

Compared to most other members of the Carnivora, domestic cats are unusually vocal, although they do not have such complex and sophisticated acoustic signals as, for example, the social mongooses (Peters and Wozencraft, 1989). A large part of the cat's repertoire consists of sounds which bring individuals closer together, and many of these were once accounted for as a by-product of domestication, until it was realized that cats can form co-operative social groups even when their contact with man is very limited (see Chapter 8). Vocal communication will only be fully understood when its role in social groups has been fully explored; at the moment we actually understand more about the function of some calls when they are directed to people than when they are used in a cat–cat context.

There are probably several reasons why cats should be so vocal. In general terms, at least four factors have been proposed for the evolution of rich, complex vocal repertoires in individual species of carnivores. These are given below.

1. Social species require more acoustic signals than those in which adults only come together to mate or to dispute territories.

2. Nocturnal species may rely on sound to take the place of visual signals.

Table 5.2. The characteristics of the vocal calls used by the cat, derived from Kiley-Worthington (1984), with additions from Moelk (1944), and the circumstances under which each is most typically used.

Name	Typical duration	Pitch (Hz)	Pitch change	Circumstances
1. Sounds produced with mouth closed				
Purr	0.5–700	25–30	—	Contact
Trill/chirrup	0.4–0.7	250–800	Slight rise	Greeting
2. Sounds produced while the mouth is opened and gradually closed				
Miaow	0.5–1.0	700–800	—	Greeting
Long miaow	0.7–1.5	700–800	Drop at end	Greeting
Female call	?0.5–1.5	?	Variable	Sexual
Mowl (male call)	?	?	Variable	Sexual
3. Sounds produced while the mouth is held open in one position				
Growl	0.5–4	100–225	—	Aggressive
Yowl	10–16	200–600	Rising	Aggressive
Snarl	0.5–0.8	225–250	—	Aggressive
Hiss	?0.5–1.0	Atonal	—	Defensive
Spit	0.01–0.02	Atonal	—	Defensive
Pain shriek	Variable	?	—	Fear/pain

3. A long period of association between mother and offspring will also produce a need for complex, age/sex-class-specific signals, some of which will be acoustic.

4. Closed habitats (e.g. forest) place a greater premium on sound communication because there visual signals are effective over much shorter distances than they are in open country.

The first three criteria certainly apply to the domestic cat, and regarding the fourth, its small size may militate against efficient long-range visual displays in all but the most featureless of habitats. However, the extent to which each, if any, of these factors applies to any individual species within the Carnivora is not well understood, so the evolution of vocal communication in the domestic cat can only be guessed at.

The vocal repertoire of adult cats

A precise figure for the number of distinct sounds produced by cats depends on the authority consulted. There could be at least two reasons why this is so. The first is that while each call may have a 'typical' form, it will not be used on all occasions by all individuals, and a divergent form of one call may sound much like a variation on another. Alternatively, some types of call, one obvious example being the miaows, may form a continuum which the various classification systems have divided up to different extents. Furthermore, many of the descriptions of calls come from domestic pets, in which the form of some calls may have been shaped or trained as part of the cat–owner relationship. For example, of the 16 patterns characterized phonetically by Mildred Moelk (1944), three were only noted from her own cat. Also no systematic description has been made of the vocalizations of the oriental breeds, which vocalize more often, and possibly have a wider repertoire, than the occidental varieties.

Eleven of the most commonly distinguished calls are given in Table 5.2. They are divided according to the position of the mouth when the sound is emitted. Two, the purr and the trill, can be produced with the mouth closed. Three are produced while the mouth is opened and closed, much as human speech is; these are two forms of the miaow, which can be further subdivided, and the female and male mating calls, which can also fall into the third category, depending on their intensity. This third group are all 'strained-intensity' calls, produced in a stereotyped way with the mouth held open in a fixed position, and often accompanied by the appearance of great tension in the facial muscles. Most, apart from the version of the female call used during courtship, are associated with aggressive or defensive behaviour.

Purring

The purr is highly characteristic of the felids, but the way that it is produced has remained something of a mystery until recently, and its biological function is still not fully understood. It is easy to hear the purring that stroking often elicits in a pet cat, and it is this sound, along with various miaows, that pets most often direct towards their owners. However, by fitting free-ranging cats with throat microphones, Kiley-Worthington (1984) found that purring occurs in a much wider range of circumstances than just cat–human interactions. Purring was not only switched on by the presence of the cat's handler, but also while the cat was nursing kittens or greeting a familiar social partner, during tactile stimulation such as rolling or rubbing, drowsy sleep, and by warm familiar environments. Conversely, a variety of strong stimuli switched off purring, including aggressive or sexual interactions, a social interaction with a cat not encountered for a long time, and on first smelling catnip, while hunting and in the presence of prey. It is difficult to summarize the various circumstances under which purring is heard, but all appear to be associated with actual cat–cat or cat–human contact, or circumstances under which cat–cat contact might be desired, such as when the cat is about to sleep. This idea has been summarized as 'contact' in Table 5.2. Perhaps the most bizarre occurrence of purring is one well known to veterinarians, who report that cats will

occasionally purr when in extreme pain; it has been surmised, without any
direct evidence, that this is the cat's attempt to reduce its own stress, much
as we can reduce the perception of pain by concentrating on pleasant
mental images.

There has been a considerable amount of speculation about the way that
the purr sound is produced, not least because it is produced continuously,
during both inhalation and exhalation, except for a brief pause at the tran-
sition between the phases of the respiration cycle. It is now known to be a
true vocalization, because it involves the vocal apparatus. The sound is due
to a sudden build up and release of pressure as the glottis is closed and then
opened, and the sudden separation of the vocal folds as a result of the
pressure build up. The glottis is moved by laryngeal muscles that contract
for 10–15 ms every 30–40 ms. These are driven in turn by a very precise,
high-frequency neural oscillator, which appears to be free-running, since
no reflex arcs to mechanoreceptors in the region of the vocal apparatus can
be detected (Frazer-Sissom *et al.*, 1991).

Miaows

These apparently commonplace vocalizations have excited a great deal of
attention, but little understanding. Every language has a representation of
this type of call:

> The English cat 'mews', the Indian cat 'myaus', the Chinese cat
> says 'mio', the Arabian cat 'naoua', and the Egyptian cat 'mau'. To
> illustrate how difficult it is to interpret the cat's language, her
> 'mew' is spelled in thirty-one different ways, five examples being
> maeow, me-ow, mieaou, mouw, and murr-raow.
>
> (Quoted in Moelk, 1944)

All, including the trill or greeting-miaow, which is often performed without
opening the mouth, are uttered in amicable social encounters, to establish
contact and incite further friendly interaction, such as play, or sharing food.
The 'chirrup' form of the trill is particularly used by the mother as a
contact-call to her kittens, which by about 21 days old are capable of
distinguishing between their own mother's call and that of another female;
both mother and kittens may also have an ultrasonic call that fulfils the
same general function.

The long miaow is a high-intensity version of the ordinary miaow, well
known to owners of cats when they do not produce food quickly enough,
but its function in cat–cat interactions is obscure. Individual cats seem to
develop different miaows for specific situations when interacting with
people, but the process by which the basic call is modified for these interactions has not been studied. It would be interesting to know whether there
is any basic pattern, to be found in most individual cats, of likely asso-

ciations between particular miaows and types of situation. For example, food begging might usually be associated with one type of miaow, a request to open a door with another. However, such a study would be complicated by the very individualistic form of the basic miaow that some cats have. Finally, some cats also have a soundless version of this call (i.e. a visual display), in which the mouth is opened and closed at the same rate as a miaow; this evidently has appeal to owners, but its role, if any, in cat–cat behaviour is unknown, although it is also performed between farm cats.

Strained-intensity calls

Both the female call, indicating readiness to mate, and the aggressive growls and wails of competing tom cats, are most often heard in the context of sexual activity. The yowl or 'caterwaul' can be emitted on its own or in combination with growling. It is probably no coincidence that the growl is one of the lowest-pitched calls used by the cat. Low-pitched sounds are generally characteristic of large animals, and so in aggressive encounters, each cat may be trying to deceive the other into believing that it is larger than it really is. The hiss is primarily a defensive sound, that proceeds to actual aggression only if the other cat presses home its attack. The other defensive threat, the spit, is the only call the cat uses that starts abruptly and at full intensity; it is probably the vocalization that is used most often to deter predators, and as such needs to be an unambiguous signal interpretable by species other than the cat.

Kitten calls and the development of adult calls

For the few days after its birth, the kitten has only two basic types of vocalization, a distress call, by which its mother is guided to attend to it, and the defensive spit. The distress call is given when the kitten is hungry, when it is trapped (e.g. if the mother inadvertently lies on top of it), or if it becomes isolated and cold (Haskins, 1979). The rates of vocalization that these circumstances produce vary with the kitten's age, as the probability and consequences of each event change. Restraint always elicits a high rate of calling in kittens up to six weeks old, but kittens exposed to cold cry less as they get older and more able to regulate their own body temperatures (Fig. 5.1). Cold elicits a call of higher frequency and shorter duration than occurs in other circumstances, and it may be that the mother can distinguish two or more types of the kitten call, and tailor her actions to suit. For example, the 'cold' call may stimulate her to retrieve a young kitten that has clung for too long to her nipple, and has dropped off well outside the nest. The effect of isolation from the mother (without cooling) peaks in the third week, presumably once the kitten has formed a mental picture of the social group in which it finds itself. Movements of both the mother and

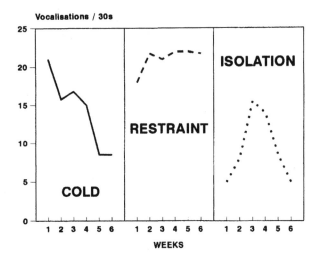

Fig. 5.1. Rates of vocalizations shown by kittens exposed to cold, restraint and isolation in each of the first six weeks of life. (Data from Haskins, 1979.)

the kitten's littermates can stimulate low levels of crying, but kittens that are huddled together and not disturbing one another hardly ever cry. Even in a strange environment, the presence of littermates reduces the level of crying; this is presumably an adaptation to prevent the kittens giving their position away immediately after their mother has moved them to a new nest-site.

The response of the mother to the normal kitten call is context dependent. If the sound comes from a location away from the nest, she is stimulated to leave the nest and investigate. If the sound comes from inside the nest, and she is herself outside, she will return to the nest and enter it. If she is inside, and with the kittens, she will usually sit down, and then lie down, allowing them to suckle. If for some reason one or more kittens are unable to locate a nipple, they may continue to cry, in which case the mother is further stimulated to shift position, jerking her abdomen so that the nipples are better exposed (Haskins, 1977). As each kitten grows, its call takes on characteristics that are individual specific, but it is not known whether the mother can recognize each member of her litter by their cries.

From the time they are a few days old, kittens can also purr, and do so particularly when they are suckling. In this situation, the sound will be transmitted to the mother and the littermates as much as by direct skin contact as through the air. The purr may signify to the mother that the kitten is receiving a satisfactory amount of food. As the kitten grows, it may associate situations of amicable contact with its own, its littermates' and its mother's purring, and carry this association through to amicable contacts in adulthood, explaining the retention of purring through life.

Up to about three to four weeks of age, a kitten in pain will emit a high-intensity version of the kitten-call, as well as spitting, but from about that time onwards the adult form of the pain shriek begins to emerge in its characteristic form. The kitten-call itself is common until the kitten is two or three months old, but then declines until it is rarely heard from kittens over five months old. Miaows first appear soon after weaning, and at that stage are more likely to be directed at a human providing food than towards other cats. At about the same time, the kittens begin to snarl at one another, particularly in contests over food, and gradually the growl and anger wail are added to the repertoire of aggressive calls. The defensive hiss is absent until the kittens are at least a month old (Brown *et al.*, 1978).

Some of the changes that take place in the characteristics of the kitten calls, such as their frequency, can be simply ascribed to the physical development of the vocal apparatus. Others, such as the duration of the call, tend to decline up to the end of week three, probably reflecting the kitten's growing independence at this stage, and therefore a reduced need to call for its mother's attention. Because kittens are virtually deaf when born, it is not surprising that the initial form of the kitten-call does not depend on auditory experience in any way. However, from about ten days onwards, normal kittens control the loudness and frequency range of their calls by listening, both to themselves and to littermates. By three weeks of age, the calls of deaf kittens tend to be distinctly louder and lower pitched than those of normal kittens (Romand and Ehret, 1984).

Visual Signals

While cats do not have any structures that are obviously specialized to produce visual signals, they appear to have a highly expressive body language that moderates a whole range of social interactions. Such signals are much more difficult to simulate than are sounds, which can be recorded and played back, or scent marks, which can be moved from place to place or even extracted and redeposited on a new surface. Their effect can only be judged from the reactions of other cats during social interactions, and as such their functions can only be inferred. Paul Leyhausen (1979) has described and categorized in great detail the body postures that are associated with agonistic behaviours, and much of the account that follows stems from his work. Other visual signals, although familiar to anyone acquainted with cats, are less well understood.

Whole-body signals

Cats, like many other mammals, change their apparent size during aggressive encounters. An attacking cat will cause its fur to stand on end (piloerection) and draw itself up to its full height; only the external ears are drawn back, presumably because these are often damaged in fights, and can subsequently become badly infected (Fig. 5.2, top right). A submissive cat will crouch on the ground, flattening its ears and coat, and generally try to look as small, and presumably as unthreatening, as possible (Fig. 5.2, bottom left). Its head is drawn into its shoulders, indicating that it is not in a position to launch an accurate biting attack, and indeed defensive aggression usually starts with a swipe of the paw. The completely submissive posture shown by the dog that rolls over and exposes its inguinal region is uncommon in the cat, at least in the context of aggressive behaviour, although Leyhausen's book contains drawings of a submissive cat that has backed into a corner and is performing a near-headstand as a submissive signal, while still swiping at its aggressor with its front paws. Leyhausen considers a third agonistic posture, the arched back, to be the result of conflicting tendencies to attack and defend (Fig. 5.2, bottom right), and has illustrated a number of intermediate postures that connect all three extreme agonistic positions (Fig. 5.2, centre). A cat in the arched-back posture often stands sideways on to its opponent, which may be another cat, a dog, or if the cat is an unsocialized feral, a human. A similar posture is a common component of play in the kitten (see Chapter 4).

The use of body language in non-agonistic encounters is less well documented. Females in oestrus go through a sequence of postures (described in Chapter 8) that includes rolling on to the back and on to the feet again. Some adult males also roll over on to their backs, apparently as a form of display, but rarely, if ever, as part of agonistic encounters. High-intensity displays of object rubbing (described below) may be preceded or followed by rolling, suggesting that some sort of transfer of scent may be taking place, but in other instances rolling can form part of social encounters. This display may therefore have a meaning outside sexual behaviour, but what this might be is not clear. It also forms part of the catnip response (see below).

Facial expressions

In addition to its body posture, a submissive cat will often deliberately avoid eye contact with its aggressor, by pointing its head to the side. Anthropomorphically, this might be interpreted as the submissive cat 'taking no notice' of the other, i.e. actively avoiding the confrontation, but the escalation of violence that ensues if the cats do make eye contact shows that the individual that is looking away is actually highly aware of the other. The ears can also be used to express submission (Fig. 5.3, centre left), by being folded sideways and downwards (as opposed to backwards rotation, which is part of the offensive posture; Fig. 5.3, top right). Ear positions can be altered much more quickly than whole-body postures, and so, while the

Fig. 5.2. The whole-body postures that signify aggression (increasing from left to right) and submission or fear (increasing from top to bottom). See text for further explanation. (Redrawn from Leyhausen, 1979, by Michael Toms.)

two dimensions of Fig. 5.3 correspond to those of the body postures in Fig. 5.2, the two do not necessarily change precisely in parallel; facial expressions can alter dramatically while the body maintains a fixed position. It would be surprising if ear and eye signals were not used in other contexts, but these have not been investigated in detail.

Tail signals

The tail is fundamentally an organ of locomotion, and cats that are jumping, climbing or galloping use their tails for balance. At slower gaits or

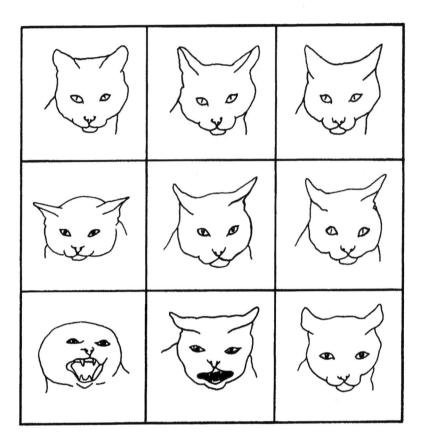

Fig. 5.3. The facial expressions that signify aggression (increasing from left to right) and submission or fear (increasing from top to bottom). See text for further explanation. (Redrawn from Leyhausen, 1979, by Michael Toms.)

while standing, the tail is available to provide signals, and is unusually expressive because the tip can be moved independently of the base. In fact, there has been very little study of the meaning of these tip-twitches, but some contexts which can give rise to the basic whole-tail positions have been documented (Table 5.3). The normal tail position is horizontal or half lowered; the most obvious signals are produced when it is raised to the vertical, probably as a greeting signal, or raised at the base and then curved progressively downwards to form the concave position (Fig. 5.2, posture immediately above the bottom right-hand corner). The defensive threat posture (Fig. 5.2, bottom right) was drawn by Leyhausen with the tail erect, but other authors have indicated that a concave tail is more typically associated with the arched back. During the preliminaries to actual fights,

Table 5.3. Tail positions used by the adult cat, and some circumstances under which each typically appears, indicating possible roles in communication.

Tail position	Circumstances
Vertical	Greeting (walking, trotting or standing) Social play; object play Sexual approach (female)*
Half-raised	Sexual approach (female)
Horizontal	Amicable approach Sexual approach (female)
Concave	Defensive aggression
Lower	Aggression
Between legs	Submission

*Any of the five tail positions may be used in the male sexual approach.
From Kiley-Worthington, 1976, and S. L. Brown, unpublished.

the tail is kept out of the way, presumably because it is susceptible to damage. The aggressive cat points its tail at the ground, and the submissive cat pulls its tail between its hind legs.

Tails can also be moved side to side, and one obvious signal produced in this way is the aggressive tail-lash that often accompanies growling. Less exaggerated movements can be observed under a wide variety of circumstances, the main exception being the stiffly raised vertical tail, but their importance in communication is unknown.

Olfactory Communication

Most carnivores lead a solitary existence for the majority of their lives, and therefore often find it difficult to predict whether visual or acoustic signals will be detected by any conspecific. Odorous secretions have the advantage that they can be left to provide information about their producer for many days after they were deposited. Because domestic cats were thought to be asocial until quite recently, most studies of their olfactory communication have been interpreted as if all cats were solitary. It is likely that odours play all kinds of roles within social groups, but so far these are not well understood. For example, it is not certain whether or not colonies of cats share a common 'colony odour' which identifies them as belonging to a particular group, in addition to odours that may be individual specific, or could give information on sex, age or status. Other social carnivores do have such systems; for example, badgers have a large pocket-shaped gland beneath their tails, the subcaudal gland, which contains a strong-smelling waxy substance. The members of a clan of badgers not only 'squat mark' this wax on to prominent objects in the clan's territory, but each badger also marks the flanks of all the other members. Most of the marking is done by the dominant male in the clan, but all the members participate to some extent, with the likely result that each attains a common odour. When badgers meet, they sniff one another's flanks and anal regions, as if to check for clan membership. If any individual is away from the group for several days, a very intense exchange of scents takes place, including a mutual mixing of subcaudal scents in which two badgers back up to one another and press their anal regions together (Gorman and Trowbridge, 1989). In common with many mammalian scents, it seems likely that much of the odour is produced by bacteria and other microflora living in the subcaudal pocket, so the mutual squat marking may allow the exchange of both the scents and the micro-organisms that produced them.

Being warm blooded, mammals are inevitably presented with the problem that their scent materials will become contaminated by micro-organisms. The scents are often produced from modified skin glands, and consist of lipid- and protein-rich materials on which a variety of bacteria and yeasts can grow, producing their own odours. In some cases, the mammal appears to regulate the microflora by producing the secretion into a specialized structure, such as the anal sacs of the domestic dog. In other cases, for example when the scent-producing material is voided on to hairy skin, there can be little control, and there will be even less once the scent mark has been deposited. How an effective communication system can work under such a set of uncertainties is still a mystery, but it has been suggested that so long as the microflora consists of a reasonably stable group of species, each animal should be able to produce an individual-specific odour that can be learned by others. The production of group-specific odours must involve not only the mixing of scented material, but also micro-organisms, which is presumably why badgers place such emphasis on exchanging the contents of their subcaudal pouches.

In addition to skin glands, the vast majority of mammals use their urine and faeces for odour communication, either relying on their intrinsic odours, or using them as carriers for glandular secretions. Urine marking has been studied extensively in domestic cats, faecal marking less so, while the use of skin glands has been described but scarcely investigated.

Urine

Tom cats are renowned for their urine spraying, an obvious and, in a pet, highly obnoxious piece of behaviour. The cat backs up to an object, and stands, tail erect and quivering, while urine is sprayed out backwards and upwards on to the object. Females will also spray urine, though not as frequently as toms, and neutering does not always prevent the development or even the onset of spraying. Both males and females tend to increase their rate of spraying when the females are in oestrus; in a study of social farm cats by Warner Passanisi (unpublished data), the females only sprayed in early oestrus and when in consort with a tom. However, in another farm group that consisted only of females, spraying was observed while the cats were hunting, which could have been due to inherited differences (members of farm-cat colonies tend to be closely related), or to a change to more male-like behaviour by females that did not have regular contact with toms.

In the mixed-sex farm colony, males sprayed most often when in consort with an oestrous female, less often but still frequently (about once every ten minutes) when hunting, and least often when in the farmyard, where they would usually be in the company of other cats. While out hunting, they directed their sprays at visually conspicuous sites, such as the sides of buildings, hay bales, fallen branches, fence posts, grass tussocks and molehills. Spraying was not seen immediately before, or during, bouts of male–male aggression, although other carnivores, domestic dogs for example, often deposit urine during competitive encounters. Spraying by males can form part of courtship behaviour (see Chapter 8).

Young cats of both sexes, adult females, neuters, and to some extent adult males, also urinate in the squatting position. For example, in one group of six females and four males, only one male regularly sprayed (Natoli, 1985). Squat urinations are usually buried as if the cat were trying to hide whatever information might be conveyed by the smell. It is thought that such urinations are primarily eliminatory in function, but squat urinations, if detected, are sniffed by both male and female cats. However, both sexes sniff sprayed urine for longer than squat urine, so it appears likely that the two types contain different information, and that sprayed urine possibly conveys more detail. The source of this extra information has not been investigated in detail, but there do seem to be chemical differences between squat and spray urine, in addition to the deliberate targeting of the latter. Some authors (e.g. Wolski, 1982) have claimed that spray urine is cloudy, and contains droplets of a viscous lipid material that might originate in the anal sacs. By comparison with most other carnivores, these are small and unspecialized, and their role in communication has scarcely been investigated. Their secretion is usually yellow-brown in colour and has a characteristic odour, and it is likely, since this is known to occur in male lions and tigers, that it could be discharged into the stream of sprayed urine. On the other hand, tom cats can produce a strong-smelling clear liquid from beneath their tails, quite distinct from the anal sac contents, and it may be this that gives the sprayed urine of intact males its pungency. The urine itself contains two unusual amino acids, felinine and cysteine-*S*-

isopentanol, that are produced in the kidneys. These are unlikely to make much impact on the odour in their own right, but their microbial break-down products might conceivably contribute to the odour of stale urine.

Information conveyed by urine marks

The amount and type of information conveyed by a urine mark has not been investigated in detail, and the only measure that has been used at all widely is the amount of time that cats spend investigating a given sample. Males generally spend rather longer at this than females, unless the female is in oestrus, when she will pay great attention to the sprayed urine of males, particularly if they are strangers. Correspondingly, males sniff oestrous female urine for longer than anoestrous. Males also appear to discriminate better between urine samples from unknown cats, cats from neighbouring groups, and cats of their own group (Fig. 5.4), although females also indicate that they make this distinction. This does not neces-sarily mean that each cat can recognize each urine scent mark as belonging to a particular individual, since in social groups there is the possibility of 'clan odours'. These odours are maintained either actively by transfer of odour materials or odour-producing microflora between individuals (as described above for badgers), or passively, for example by the sharing of food sources with characteristic odours within, but not between, groups.

Urine is not only sniffed, it can also induce the Flehmen or 'gape' re-action in which the vomeronasal organ is activated (see Chapter 2), although the extent to which this additional form of olfactory inspection is used seems to vary considerably from cat to cat and is possibly commonest towards the scent mark of a strange male. While all cats will initially sniff a urine scent mark, many follow this with bouts of Flehmen, in which the tongue is moved rhythmically along the roof of the mouth, and then the head is raised, the mouth is opened, a posture which is maintained for up to a quarter of a minute. Head shaking may occur between bouts of Flehmen. Sniffing at a distance, when the cat behaves as if it is then attracted by the odour of the urine, is only common when the urine sample is fresh; older marks are often detected because cats tend to inspect the prominent objects on which they were deposited. Prolonged sniffing of old scent marks can result in the mark being moistened by the cat's nose, and Flehmen will sometimes follow, although less commonly than to fresh marks.

Before it became clear that scent marks of all kinds might play a role in regulating social behaviour, it was thought that the primary function of spray-urination was to enable hunting cats to space themselves out. The hypothesis was that by avoiding areas that had been recently hunted by others, each cat could maximize its chances of encountering undisturbed prey, and minimize the probability of an encounter with another, possibly

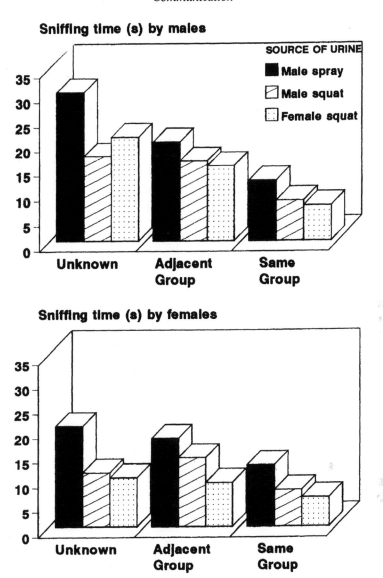

Fig. 5.4. The amount of investigation induced by three types of urine from three classes of cat, in male (top) and female farm cats (bottom). (Data from Passanisi and Macdonald, 1990a.)

hostile, cat. The odour of a scent mark changes as it ages, due to a combination of differential evaporation of volatiles and the production of new smells by micro-organisms, and could be used to indicate the time since a cat last hunted that area. Such an argument gives more advantage to the receiver of the message than to the emitter, and modern theories of communication tend to emphasize the idea that emitters of signals are attempting to manipulate the behaviour of receivers rather than vice versa. For example, a tom cat that did not spray might gain an advantage over its rivals by inducing them to hunt in areas that he had recently disturbed. Certainly, foraging cats do spend a great deal of time investigating scent marks, but their spray-urinations tend to be concentrated in the core of their hunting territory; at the edges of their home ranges, their rate of urination tends to drop, which would not be predicted by the time-sharing hypothesis. Individual cats may indeed make use of the fresh urine marks to inform themselves of the likely whereabouts of others, but it is likely that these marks also confer some advantage on their emitter. What this advantage might be is still uncertain, but in the case of males some function connected to the mating system cannot be ruled out.

Faeces

Many carnivores use their faeces, sometimes with glandular secretions added, to disseminate olfactory information. For example, European otters deposit piles of faeces, known as 'spraints', at nose height, throughout their home ranges on the tops of prominent objects such as large rocks, where they can be easily detected by other otters. The role, if any, of faecal odours in cat–cat communication is not yet certain. Farm cats usually have communal latrine sites, which tend to be in areas with a loose substratum, such as turned soil, gravel, sand or hay. Burying of faeces is common at these sites, although some stools are always exposed, either because no attempt has been made to bury them or because in burying one defaecation another has been uncovered. The significance of such communal sites may rest on the scarcity of areas with suitable substratum, rather than being connected in any way to communication. Away from the farmyard, piles of faeces can occasionally be found on exposed, conspicuous sites. If the primary function of covering is one of hygiene, then it may not be necessary away from the centre of a communal territory, but the use of obvious sites cannot be explained in this way, and suggests some sort of communicative function.

Scratching

Within a cat's home range, certain objects are marked repeatedly by being scratched with the claws of the front feet. Favoured sites are usually vertical wooden surfaces, such as the trunk of a tree or the side of a shed, and these can become deeply grooved if used over a long period, providing a potential visual signal. Fragments of claw and claw sheath can often be found embedded in these grooves, giving the impression that the primary purpose of the scratching is to condition the claws. However, if this were the case, then presumably there would be at least some need to scratch with the claws of the hind feet as with those of the front. The scratched sites presumably act as combined visual and scent signals, the latter derived from the sebaceous glands of the feet. The precise role of these signals, and their relative importance compared to other scent marks, such as urine and faeces, is unknown. Feral cats perform claw sharpening more often in the presence of conspecifics than when alone, suggesting that the action itself may serve as a visual gesture of dominance (Turner, 1988).

Skin glands

Cats have a variety of specialized skin glands, several of which appear to be communicatory in function. There is a large submandibular gland beneath the chin, in the area where most other carnivores have a tuft of whiskers, but it produces no obvious secretion and most cat owners probably never know of its existence unless it becomes infected. There are perioral glands at the corners of the mouth, temporal glands on each side of the forehead, and diffuse clumps of sebaceous glands along the tail, collectively known as the caudal glands (Wolski, 1982). Other skin glands, at the base of the tail, enlarge as the cat matures, particularly in intact males, and may be the source of the clear liquid mentioned above. The cheeks and ears may also produce odorous secretions, and the mutual flank-rubbing display performed by familiar cats suggests that this area also may be endowed with scent-producing structures.

The secretions of the submandibular, cheek and perioral glands can be deposited as scent marks on prominent objects at head height, such as twigs that project over frequently used paths. Known as bunting, this marking behaviour can be performed in isolation, or following Flehmen towards a urine mark or head mark, and may itself be followed by spraying if the marker is a tom. The precise form that head rubbing takes seems to depend at least partly on the topography of the object being rubbed. Cheek rubbing is performed along a line from the corner of the mouth to the ear. Higher objects may be marked on their undersurface using the forehead and ears; objects near the ground may be marked first with the underside of the chin, then with the side of the throat (Verberne and De Boer, 1976). Inanimate objects may also be rubbed with the flanks and the tail, although these areas of the body are more commonly used in cat–cat rubbing.

If all the possible combinations of glands produced different scents, a complex set of messages could be available, but there is no evidence for this, and it seems more likely that many of the glands produce essentially similar odours, and can thus be used interchangeably. Solitary cats pay a great deal of attention to the sites of previous head marks when they are patrolling their home ranges, but the information contained in these marks has not been elucidated. In one group of farm cats, Warner Passanisi found that females increased their rate of head marking during oestrus, while males showed a similar increase when consorting with a female. Males would occasionally head mark away from their core area, but females never did. Females may spend longer than males in investigating head marks in general, but males pay particular attention to head marks left by oestrous females.

Allorubbing

The mutual head, flank and tail rubbing of cats will be discussed further as an important component of social behaviour. Its role in scent communication is unclear, and it is possible, though not likely, that it is a purely visual and tactile display. Within cat societies that contain breeding males, females and kittens, the initiation of rubbing is highly asymmetrical, but it is not known whether the purpose of such encounters is to deposit, pick up or share odours. When cats sniff one another, they tend to concentrate on the facial and perineal areas, suggesting that any materials deposited on the flanks and tail by allorubbing may be less important as social cues than individual odours produced by the head and anal glands, and around the genitals.

Catnip

The response of some cats to the herb catnip (*Nepeta cataria*), and other plants containing compounds similar to nepetalactone, the most active compound from catnip, has been exploited for many years by the makers of cat toys, but the behavioural significance of the response is still unclear. It can be assumed that the aroma of catnip conveys some message to cats, but no adaptive significance for this has ever been elucidated, and it must be assumed that the occurrence of the triggering compounds in a few types of plant is an evolutionary accident. The primary part of the reaction is an intense combination of face rubbing and body rolling, and can be elicited in males and females, both neutered and intact. Thus, while the rolling pattern is very similar to the same behaviour when it forms a part of female oestrus behaviour, the whole catnip response does not appear to be directly related to any sexual response, nor is it motivationally connected to hunting or aggressive behaviour (Palen and Goddard, 1966). The response is inherited as a dominant autosomal gene, and therefore not all cats will show it. Some other felids, again of both sexes, show a similar response, among them lions and jaguars, but not tigers or bobcats.

Feeding Behaviour

For some aspects of the behaviour of the domestic cat, it can be difficult to define a precise function for each of the behaviour patterns, because the circumstances under which they evolved may not be readily apparent today. For example, in a later chapter it will be argued that the social behaviour of cats has been substantially altered by domestication, even though semi-independent groups of cats can integrate their behaviour in an apparently functional way. However, the function of feeding behaviour is adequate nutrition, and there is no evidence that the nutritional requirements of the cat have changed during domestication; the machinery that supplies the body with energy and the building blocks for growth and repair are essentially those of a wild felid. Moreover, the whole cat family is nutritionally peculiar in a number of respects, most of which can be directly related to their carnivorous lifestyle. Of all the Carnivora, the felids are the most specialized meat-eaters; despite the name of the group, many of its members are actually omnivores, including the bears, and some, for example the pandas, have secondarily reverted to plant eating. Thus any account of feeding behaviour must logically start with a description of the goal of that behaviour, the fulfilment of nutritional requirements.

Nutrition

A meat diet is rich in protein and fat, but low in carbohydrate compared to a vegetarian diet. There are also many qualitative differences within these broad categories, and omnivores have a wide range of biochemical mechanisms for converting nutrients from both plant and animal sources into forms that they can use directly; since omnivores are themselves animals, the conversions tend to be more complex when plants are the source. To take a simple example, the visual pigment in the retina, rhodopsin, is partly formed from retinol, also known as vitamin A. A similar group of compounds, the carotenes, are commonly found as yellow and orange pigments in fruits and vegetables, and most omnivores use these as precursors of vitamin A. Cats do not have this ability, and therefore need to obtain all their vitamin A as vitamin A itself, from animal sources. Strict specialization as meat-eaters appears to have resulted in the loss of certain metabolic functions, which now means that cats are unable to obtain a balanced diet from plant materials alone. Further examples of these restrictions will be discussed as each group of nutrients is considered. Cat nutrition is described in considerably more detail by MacDonald *et al.* (1984), the National Research Council (1986) and Edney (1988).

Protein

Cats require a much higher proportion of protein in their diet than almost any other mammal; the minimum level for body maintenance in the adult is about 12% (of dry weight of food) and growing kittens need about 18%. The main reason for this unusually high requirement is that the cat's liver contains highly active *N*-catabolic (protein-degrading) enzymes which cannot be shut off completely. As a result, both the adult and the kitten automatically take a certain amount of energy from the protein part of their diet, whereas the majority of mammals have the ability to direct almost all their dietary protein into growth and body maintenance. For example, although the protein requirements of the fox are very similar to those of the cat, those of the domestic dog (4% adult, 12% puppy) are much more typical of an omnivore. Note that the puppy requires three times as much as the adult dog, but the kitten requires only one-and-a-half times as much as the adult cat, reflecting the cat's underlying need for protein as an energy source at all stages of life. Presumably, since the natural foods of cats and other specialized carnivores are always protein rich, they have not suffered from losing the ability to conserve protein when it is in short supply.

All of these figures only apply to protein that contains the right blend of amino acids. The list of amino acids that are essential in the diet does not vary greatly between one mammal and another; in the cat these include arginine, histidine, lysine, leucine, isoleucine, valine, methionine, threonine, tryptophan and asparagine. The cat is, however, much more susceptible to a deficiency of arginine than are other mammals. A single arginine-deficient meal can lead to ammonia intoxication, which includes emesis and lethargy among its immediate symptoms. The missing arginine is required as an intermediate in the urea cycle, which is essential for the excretion of nitrogen, and when it is absent that nitrogen accumulates as ammonia in the blood. The reason that the cat is so susceptible is that, following each meal, they deplete their levels of ornithine, another urea cycle intermediate that can substitute for arginine if the latter is in short supply. The depletion of ornithine prevents the sudden surge of amino acids in the bloodstream that follows the next meal from being broken down immediately to provide energy. Cats are virtually unable to make ornithine from any other source, and so require arginine at every meal to restart the urea cycle.

Cats also need unusually large amounts of sulphur-containing amino acids, such as cysteine and methionine, and it is these that are often the most limiting in normal foods. Because these amino acids are used in large quantities for making hair, the cat's thick coat has been suggested as one reason for this heavy requirement. Cysteine is also used for the synthesis of felinine, the unusual branched-chain amino acid found in cats' urine (see Chapter 5), but the amount made is not great, and cannot account for more than a small fraction of the cysteine requirement. Cats also need a β-amino acid, taurine (2-aminoethane sulphonic acid), which is not found in proteins, but is normally biosynthesized from cysteine. The enzyme responsible for this biosynthesis has very low activity in cats, resulting in a dietary requirement. Long-term taurine deficiency can bring about a degenerative disease of the heart muscles (dilated cardiomyopathy), retinal degradation, and can also lead to poor reproductive performance.

Fat

In general, fat has three functions in the diet, those of providing energy, acting as a solvent for some vitamins, and supplying essential fatty acids. Unsurprisingly for a carnivore, cats can make very efficient use of very high levels of fat in their diet, and moreover can digest fats that come from a wide variety of sources, including plant oils. One common adaptation to a carnivorous way of life appears to be the loss of one or more enzymes that produce arachidonic acid, itself a precursor of prostaglandins and other compounds essential to reproduction; such changes are found in animals as diverse as the lion, the turbot and the mosquito. This means, for example, that without arachidonate in the diet queen cats will not come into oestrus. Omnivores are capable of converting linoleic acid, a common constituent of plant oils, into arachidonic acid, and do not therefore require the latter in the diet. The only known precursor of plant origin that cats can use is found in evening primrose oil.

Carbohydrate

Provided their diet contains both protein and triglycerides, cats do not need any carbohydrate. Many commercial cat foods, particularly dry formulations, contain significant amounts of carbohydrate, and cats are perfectly capable of digesting starches and most sugars, including sucrose and glucose. Like most carnivores they are incapable of digesting cellulose. One adaptation to their carnivorous habits is a reduced glucokinase activity in the liver. In herbivores and omnivores (including the domestic dog), this enzyme copes with the sudden increase in glucose that follows an easily assimilated carbohydrate-rich meal. No such surge follows meat meals, so this enzyme is redundant in specialized carnivores. Cats also have an inefficient lactase, which means that they are only able to digest low levels of lactose. Any excess passes to the hind-gut where it can be fermented by bacteria, sometimes causing diarrhoea. This accounts for some cats' intolerance of milk. Cats find milk fats highly palatable, but the presence of quantities of lactose in cow's milk means that this is not a suitable food for all cats.

Minerals and vitamins

The cat's mineral requirements are not much different from those of other mammals. In the wild, they must consume at least some of the bones of their prey, because meat alone is deficient in calcium and phosphorus. Some cats appear to be intolerant of high levels of magnesium in their diets; crystals of the magnesium salt struvite can form in the bladder giving rise to feline urological syndrome. One of the advantages of a carnivorous lifestyle is that all food is comparatively rich in sodium, essential for nervous and excretory functions. Most plants contain much less sodium than herbivores require, and so most mammals have a highly sensitive taste for salt, and some will travel long distances to salt-licks to obtain sufficient of this important mineral. Both the cat and dog are rather insensitive to the taste of salt, as will be described below, reflecting the likelihood that their diet will automatically contain enough sodium.

Cats have a few peculiar requirements for vitamins. Like the majority of mammals, they do not require vitamin C, but, as mentioned in the introduction to this section, they do have a requirement for vitamin A that cannot be met from plant-type precursors. Niacin is also essential, because the cat can make very little of this vitamin from tryptophan (a competing biochemical pathway is very active). Thiamine is also required in abnormally large amounts; however, since this vitamin is mainly used in carbohydrate conversions, less is needed when the primary sources of energy are fat and protein. The raw flesh of fish contains thiaminases, enzymes which destroy thiamine, so a diet of uncooked fish can result in thiamine deficiency. Presumably the fish-eating cats of India (*Felis viverrina*) have some mechanism for overcoming this problem.

Digestion and water relations

The gut of the cat follows the typical carnivore pattern; the fore-gut is emphasized, the hind-gut is reduced, and the colon and rectum are not obviously differentiated from one another. The caecum is small. Overall the gut length is only four times the body length, reflecting the high digestibility of much of the cat's food (Edney, 1988).

Meat has high water and protein contents compared to many other foods, and, unless heat stressed, cats can live on the moisture from meat (or commercial canned cat foods) alone, without drinking. The kidneys are capable of producing much higher concentrations of urea than those of man (2000 mM compared to < 800 mM), which is presumably an adaptation to the arid habitat of the ancestral populations of *lybica*. For this reason, some cats drink only occasionally, although they must have water available if they are fed on dry or semi-moist cat foods. Whole milk, although a popular addition to many companion cats' diets, is less suitable as a source of water, partly because it contains substantial amounts of protein, fat and carbohydrates and is therefore a food in its own right, and partly because some cats are intolerant to the milk sugar lactose. The attractiveness of milk seems to lie in its fat component; witness many cats' predelictions for butter and cheese.

Taste

Cats undoubtedly use both the sense of taste and the sense of smell in selecting food, and physical characteristics such as texture and temperature can also be important. Olfaction has been discussed in Chapter 2, but its precise role in food selection has received little study. Cats probably use smell as a secondary cue in detecting their prey; hearing and sight are probably more important, as will be discussed in Chapter 7. Many owners will have noted their cats sniffing at an unfamiliar food before eating, but the whole flavour of a food, made up of a combination of its smell and its taste, is probably more important than odour, taste or texture alone in sustaining actual eating. For example, exposure to odour alone will not overcome neophobia to a strange-smelling food, whereas actually eating that food will do so rapidly. Some experiments with dogs have indicated that meaty odours alone will not sustain interest in an otherwise bland food, and it is likely that cats are also not easily misled in this way (Bradshaw, 1991).

The sense of taste has been studied in some detail in several domesticated animals, including the cat. Human taste sensations can be easily measured by asking subjects to rate their responses to standardized tastants; such reporting is of course unavailable when the subjects are

Table 6.1. Summary of the major groups of taste neurones in the facial nerve (geniculate ganglion) of the cat, with their equivalents in the domestic dog, the rat and man included for comparison.

Neural group	Cat	Dog	Rat	Man (psychophysical equivalent)
Amino acid (cat type)	M	M	(m)*	Sweet/bitter
Acid (cat type)	M	M	—	Sour
Acid (rat type)	—	—	M	—
Nucleotide	M	m	—	(Some 'bitter')
Salt	—	—	M	Salty

Adapted from Boudreau, 1989. M = major group, m = minor group, (m) = imperfectly defined.

*The amino acid groups of the rat are qualitatively different from those of the cat and the dog, and are predominantly located in the glossopharyngeal nerve.

animals, and so most of our information on animal taste responses comes from neurophysiological investigations (Boudreau, 1989). There are four cranial nerves that convey information on taste from the mouth to the brain, but of these only the facial nerve has been studied in detail. A few accounts describe the basic properties of taste-sensitive neurones in the other three (the glossopharyngeal, the vagus and the trigeminal), but these have tended to focus on their responses to just a few simple compounds. Several distinct taste systems have been identified in the facial nerve (Table 6.1), and the responsiveness of each has been established to a wide spectrum of taste compounds and nutrients, so that it is possible to make comparisons between the cat's abilities to taste particular types of compound, and the nutrients that it is likely to encounter in its food.

Amino acid units

The taste-buds are found both on fungiform papillae on the front, edges and top surface of the tongue, and also in four to six large vallate (cup-shaped) papillae at the back of the tongue. The commonest type of taste unit in the facial nerve responds primarily to amino acids, and most of these originate in taste-buds at the tip of the tongue. These units, in common with most of the others that will be described, produce a low rate of discharge spontaneously, and it is therefore possible for particular amino acids to cause either an increase and/or a decrease in the rate of discharge. By analogy with human taste descriptions, it seems likely that increases can produce one taste sensation, while decreases produce another, quite

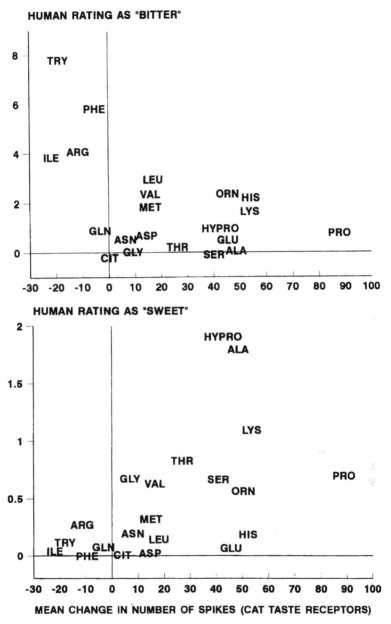

Fig. 6.1. Responses of the cat's amino acid sensitive taste units to a series of 21 amino acids. Some amino acids excited these units (produced more spikes; positive values) and some inhibited them (inhibited the spontaneous production of spikes; negative values). Each is plotted against the psychophysical values reported by human tasters of the same solutions, rating 'bitter' (upper graph) and 'sweet' (lower graph). (Redrawn from Boudreau *et al.*, 1981.)

distinct sensation. Some amino acids trigger rapid rates of discharge in the amino acid units (Fig. 6.1), while others (mainly those with hydrophobic side-chains) inhibit the discharge. The former category are generally described as 'sweet' by human tasters, and the latter as 'bitter', but of course this does not necessarily mean that cats experience the same subjective sensations that we do. It is known, however, that cats spontaneously prefer solutions of the 'sweet' amino acids, particularly proline, histidine and lysine, to plain water, and reject those in the 'bitter' group (White and Boudreau, 1975), so the analogy with our sweet taste at least may be a reasonable one. It is interesting that as far as their response to amino acids is concerned, these receptors are much more like those found in man than those of the rat. For example, L-arginine, which inhibits the cat units and tastes bitter to us, is highly stimulatory to the equivalent receptors on the tongue of the rat.

The same taste units also give a rather weak response to both common salt and potassium chloride, but cats seem to be much less sensitive to both these compounds than the majority of mammals are. Rats, for example, have a large population of highly salt-sensitive units in their facial nerves, for which no analogous taste-buds have been found in the cat. However, many mammals also have a second population of salt-sensitive cells in the glossopharyngeal nerve, and since this nerve has not been investigated in detail in the cat the possibility of a salt-specific taste cannot be ruled out.

The labelling of these units as 'sweet' automatically suggests that they should respond to sugars, and their equivalents in the domestic dog do indeed discharge vigorously when stimulated with a whole range of mono- and disaccharides, and even some artificial sweeteners. In the cat, neither this nor any of the other taste systems identified so far will respond to sugars presented at any behaviourally meaningful concentration. Cats will in fact drink solutions of sucrose in water as if they consisted of water alone, and if thirsty will go on drinking, with potentially disastrous consequences for their water balance (Carpenter, 1956). On the other hand, they prefer sweetened to plain milk, which could be explained either by their textures being slightly different, or because the taste of the milk itself is altered by the addition of sugar; the latter would be impossible for us to detect ourselves, because the sensation of sweetness would mask out any more subtle changes.

The main function of these units is presumably to give an impression of the amino acid profile of each prey item, or even of individual tissues within the larger species caught. Cats' well-known dislike of carrion may be accounted for by the accumulation of monophosphate nucleotides in many tissues after death; these compounds inhibit the amino acid units, and are actively rejected in the same way that the inhibitory amino acids are. Dogs, which eat carrion readily, have slightly different amino acid taste units, which are inhibited by only a very few compounds.

Acid units

Our 'sour' taste is mirrored in the acid units of the cat. These respond to a whole range of strong and weak acids, such as nucleotide triphosphates, histidine, histidine dipeptides, and protonated imidazoles, as well as carboxylic and phosphoric acids. Some of the low-molecular-weight acids are preferred by cats, but medium-chain fatty acids, such as octanoic acid, are strongly disliked by cats. Even when these acids are presented as esters, enzymes in the saliva generate enough of the free acids to cause rejection (MacDonald *et al.*, 1985). Coconut oil, which is bland tasting to us, must have a strong taste as far as cats are concerned, because it contains compounds which include medium-chain fatty acid moieties. This might explain some cats' dislike of hand creams and detergents that contain coconut and other plant oils, but its broader biological significance is obscure.

Raw meat contains only low concentrations of free fatty acids, and is generally rather acidic in any case (pH 5.5–7.0), which means that many acids will not be in the ionized form necessary for the triggering of the taste response. Most of the stimulation of these units may come from the histidine dipeptides present. A few amino acids, most notably the sulphur amino acids L-cysteine and L-taurine, also trigger the acid units. At first sight it might seem that the response to taurine is an adaptation to the cat's unusual requirement for this compound (see above), but there is no behavioural evidence that cats reject diets deficient in taurine either immediately or even after repeated exposure. Furthermore, the acid units of the dog, which can make all its own taurine from cysteine, are more sensitive to taurine than those of the cat. This response, though measurable, may not have any particular biological significance beyond adding another component to the overall perception of flavour.

'X-units'

The remaining taste groups in the facial nerve are less well understood, but are all characterized by long latencies between stimulation and response, and irregular bursts of spontaneous discharge, rather than single spikes. All respond to nucleotide di- and tri-phosphates (Table 6.1), but rather poorly defined subgroups also respond to quinine (hence labelled the 'bitter' subgroup), tannic, malic and phytic acids, and alkaloids. Cats are certainly very sensitive to quinine, rejecting it in solution at a thousand times the dilution rejected by rabbits or hamsters. The behavioural significance of the X-units has not been established, but they do seem to be associated with carnivorous habits, because although they are uncommon or absent in herbivores, analogous systems are present in puffer fish and blood-sucking arthropods.

Meal Patterning

Cats are opportunistic feeders, and will vary their patterns of activity to take account of the availability of food, whether hunted, scavenged or provided by a human owner. Feeding and drinking behaviour is accordingly not coupled to any daily rhythm, although more long-term rhythms do control body weight (see Chapter 3). If palatable food is available all the time in a form that does not 'stale', for example a commercial dry food, cats will take small meals (typically 12–20) throughout both the day and night (Mugford, 1977). The meals tend to be slightly larger during daylight hours than at night. The way that meal size and the interval between meals is determined is not fully understood, but over the course of a few hours the majority of cats are able to regulate their calorie intakes to match their requirements, usually by decreasing the content of each meal rather than by reducing the number of meals. This contrasts with the situation in the majority of domestic dogs, which will over-eat if fed *ad lib*. While this represents an underlying mechanism for regulating meals, house cats are of course capable of adapting to meal patterns imposed on them by their owners, and many pets are provided with food only two or three times each day, although the food may not all be consumed immediately, increasing the number of actual meals to, say, five or six. It has been shown that as a cat has to do more work to obtain a meal, it tends to take fewer, larger meals (Kaufman *et al.*, 1980), presumably reflecting the situation of a cat hunting medium-size prey that requires substantial effort to be caught.

Learning About Food

Because cats are specialized carnivores, it is entirely feasible that they could rely more on inherited notions of what is edible and what is not, than on the complex mechanisms of food learning used by omnivores. In fact, as will be described in more detail in the following chapter, feral cats eat a wide variety of prey, which varies with the seasons and can include mice, voles, rabbits, rats, birds, lizards and insects as well as food scavenged from human sources. House cats eat a range of commercial products, as well as various processed and cooked foods primarily intended for human consumption. It seems hardly likely that information on the nutritional content of each one of these could be genetically programmed, and in recent years it has been shown that cats can indeed learn a great deal about the consequences of eating particular foods.

To put these abilities into context, it is worth making some comparisons between species with different feeding habits. Eating generally depends on two criteria being satisfied, that the potential food is indeed food, and that if eaten it will satisfy some internal need state (Rozin, 1976). For some animals, filter-feeders for example, internal states seem relatively un-important, and feeding always takes place if recognizable food is available. Other species feed on a few nutritionally interchangeable foods, one example being the lions of the African savannah, and may therefore have only one need state, that of general hunger, since a single meal will satisfy all their nutritional needs; only vitamin-deficient zebras can produce vitamin-deficient lions. The greatest problems are faced by the omnivores, which have to choose between a very large number of possible foods, each of which may offer a different blend of nutrients, some matching current needs and some not. The Norway rat is, after man, the best studied of the omnivores, and can be used as a basis for comparison with the cat.

Rats behave in a relatively inflexible way towards two key nutrients, water and sodium. Hunger for water, so distinctive that we call it thirst, arouses a search for fluids, and these are, in animals, probably detected by their textural and taste properties. Subsequently, of course, learning takes over to guide the thirsty animals towards likely sources of liquids. Rats also have a built-in preference for salty tastes, which is only activated when there is an internal need for sodium. This preference appears the first time that a rat becomes sodium deficient, and all the evidence indicates that it is innate, although, as for water, foods which contain salt can be, and pre-sumably usually are, learned. Hunger for calories, usually called just plain hunger, is also innate, but since calories can come at different levels and proportions from fat, protein or carbohydrate in different foods, experi-ence plays a major role in modifying meal sizes to take account of their calorie content. There is no simple system that could specify which foods are adequate sources of calories, and which are not, although there are crude links, for example between sweet tastes and the probability of high energy contents. In rats the recognition of the calorie content of a meal is primarily determined by the level of glucose in the blood following that meal; an association between the flavour of that meal, and its energy content, is learned, and used subsequently in regulating the intake of that food.

Intakes of all other nutrients, and components of foods that should be avoided, is based on learning mechanisms. It seems unlikely that a rat could carry the equivalent of a table of nutrient requirements in its head, hard-wired to their flavour characteristics, not to mention all the toxins that it is ever likely to encounter. For many years it was unclear how learned mechanisms could possibly play a part, because experimental psychologists believed that associations could only be learned between events that took place simultaneously or very nearly so. It is now clear that, whatever the restrictions on other types of learning (see Chapter 3), learning about foods can and almost invariably does take place between sensory stimuli and physiological consequences that occur up to several hours later. For this type of learning, it is important that the sensory stimuli are biologically relevant. Taste, and possibly odour, are readily learned, whereas visual or auditory stimuli are not. The original experiments in this area were done with foods that contained sublethal doses of poisons, which are subsequently avoided because of their unpleasant consequences. However, it is now clear that very similar learning mechanisms allow rats to correct deficiencies in their diet, by avoiding those foods that have induced the deficiency, and eating more of those that have not. For example, a thiamine-deficient diet rapidly becomes aversive; when it is presented on a subsequent occasion, the rats approach it eagerly, sniff it, but then do not eat it. Foods with the same smell as the deficient food are also avoided, even those that are not themselves deficient, and this avoidance persists after thiamine balance is restored. In fact, the rats behave towards vitamin-deficient diets just as if they were slow poisons. The general principle that lies behind this sort of learning seems to be 'avoid the most relevant stimuli associated with disadvantageous gastrointestinal consequences'. Within the mixed diet typical of an opportunist omnivore like the rat, the association is generally established between the most recently eaten new food and the unpleasant consequence that followed, on the assumption that foods that have proved themselves to be safe in the past are much less likely to have contributed to the malaise than one of which the rat has no prior experience. A tendency to sample only one new food at a time, separated by intervals of 30 minutes or more, further increases the likelihood that such learning will be accurate.

Rats can therefore learn from their mistakes. The number of mistakes that they make is kept to a minimum by neophobia, a distrust of new foods in general, which can become stronger following several unpleasant experiences. However, while they are actually in the nutrient-deficient state, rats faced with a choice between the familiar food that they now know is deficient, and an unfamiliar food, will often choose the latter. Proof that rats can also learn about some of the beneficial consequences of eating particular foods over and above their energy content has been harder to obtain, probably because the changes in preference that result are much less dramatic than those associated with deficiencies. For example, it has recently been demonstrated that rats, even when they are not deficient in protein, can learn which one of a pair of foods has the higher protein content, and will subsequently prefer the high-protein food when they do need protein. When the need is satisfied, the preference is no longer expressed, at least until the same need arises again. Such preferences are therefore termed state specific, to distinguish them from the aversions to deficient foods, which are expressed whether or not the rat is actually deficient in that particular nutrient.

Despite the relatively restricted range of foods accepted as such by cats, compared to the catholic tastes of rats, most of the mechanisms described above occur in both species, although the cat has been investigated far less thoroughly.

Learning about flavours

Cats are generally quite ready to accept novel foods. Laboratory rats are much less neophobic than are their wild counterparts, and an attenuation of neophobia may be a common consequence of domestication, as man takes an increasing role in the selection of food. The extent to which cats are distrustful of new foodstuffs may depend in part on the types of food that their mother introduced them to when they were kittens; it appears that food fixations can sometimes develop in kittens under circumstances that are not yet fully understood. The 'personality' of the kitten may also be important, because a tendency to neophobia may be linked to nervousness in general. Kittens raised on commercially prepared foods generally show the opposite trend, that is they prefer new types to the brand they have been raised on (Mugford, 1977).

It can sometimes prove difficult to introduce new foods into the diet of adult cats as well, and in some instances neophobia is the major cause of refusal, although it is often the case that the cat simply finds the food unpalatable. Depending on the individual cat, neophobia can be expressed

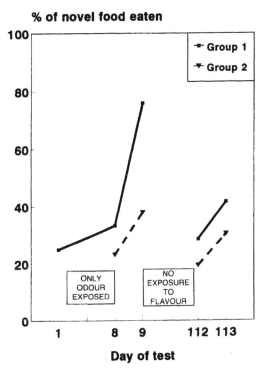

% of novel food eaten

Group 1
Group 2

ONLY ODOUR EXPOSED

NO EXPOSURE TO FLAVOUR

Day of test

Fig. 6.2. Percentages of novel (artificially flavoured) canned cat food eaten by cats presented with choices between that food and an unflavoured, familiar food. The two groups of cats were treated identically, except that only group 1 was tested on day 1. On every day between day 1 and day 8 all the cats were exposed to the odour of the unfamiliar food, but were not allowed to eat it. This exposure to the odour appeared to have little effect on preference, but by the second test both groups ate more of the novel food, and on the third test group 1 preferred the novel food. For the next three months neither group had any exposure to the novel food, after which both groups showed preferences very similar to those they had shown originally, suggesting that they were behaving in a naive way again. (Data from Bradshaw, 1986.)

Table 6.2. The influence of meal-to-meal variety on the caloric intake of cats. Three meals of nutritionally complete canned foods were offered each day to groups of cats; when each meal consisted of the same food the effect of their relative palatabilities on intake is readily apparent. When the three foods were alternated at each meal, the total calorie intake for the day was higher even than that of the most palatable single food.

Food	Intake per cat (kcal/day)
Whiskas–Whiskas–Whiskas	401
Sam–Sam–Sam	287
Kitekat–Kitekat–Kitekat	236
Whiskas/Sam/Kitekat (all possible orders)	452

From Mugford, 1977.

merely as an initial bout of sniffing or hesitation before eating, as a reduction in the amount eaten on the first encounter, or as a complete refusal to eat the new food. The second of these possibilities can result in eventual acceptance of the new food, once the consequences of eating and digesting it have been assessed, even though the cat may seem initially reluctant to eat it. It is known from other species, and seems to be true of the cat, that the circumstances under which the new food is presented can also affect the expression of neophobia. The probability of a refusal can potentially be enhanced by almost any unusual circumstance, such as a change to an unfamiliar feeding bowl, or the presence of a strange person. Exposure to the smell of the new food alone has little or no effect on how much of it is eaten on the first occasion it is presented as a meal; one or two actual meals have to be eaten before the cat's true preference is revealed (Fig. 6.2). To keep the new food in the cat's repertoire, it seems to be necessary for it to be presented at least once every few weeks, because it has been found that cats reacted neophobically towards a food after an interval of three months from the first time that they had learned that it was safe (Bradshaw, 1986).

Within their normal repertoire of foods, cats are known to appreciate variety (Table 6.2). This is perhaps most dramatically demonstrated in kittens that have been kept on a single batch of a commercial product for many weeks, as part of a trial to confirm that product's nutritional suitability. These kittens will prefer to eat almost any alternative product on the first time it is offered side by side with the diet they have been reared on, but this effect is generally transient, and the relative acceptabilities of the two diets tend to stabilize at the expected level (based on preferences recorded from cats that have not had a predominance of either food in their background diet) within a few days. In adult cats a similar effect can be demonstrated after only six days' feeding on a single product. On the

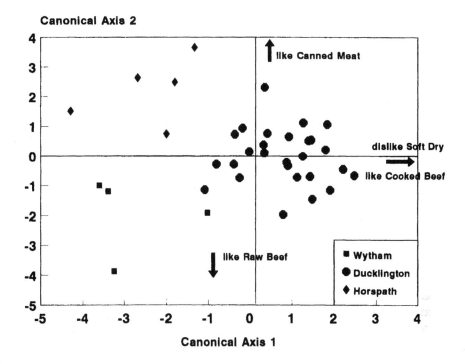

Fig. 6.3. A two-dimensional representation of the most consistent differences in food preference between cats at three farms (derived from Canonical Discriminant Analysis of 11 preference tests). Each point represents an individual cat. The major trends in 'likings' for four of the six foods tested are indicated by the arrows. The Ducklington cats were fed with the soft dry food by their farmer; the Wytham cats were given some commercial canned meat-based products by a neighbour. The Horspath group regularly fed on waste food from restaurants which contained cooked meat, including beef, but were never given any proprietary cat foods. Each group of cats therefore shows the lowest preference for the food it had most access to.

other hand, such preferences for variety can disappear if the feeding environment is changed, when the cat may prefer to eat its most familiar food (Thorne, 1982). A preference for variety is not confined to house cats, nor to those deliberately kept on a monotonous diet, because David Macdonald (of Oxford University) and I have found that farm cats show a reduced preference for types of food that form a major, but not exclusive, part of their background diet (Fig. 6.3).

Learning about toxins

The undoubted survival value of rapid learning about foods that contain toxic substances applies just as much to the domestic cat as to the rat.

Stomach upsets of all kinds can alter food preferences for some considerable period. For example, cats can develop aversions to foods containing large amounts of sugars, because they are unable to digest these in high concentrations. In one study, six hours' exposure to sucrose solution resulted in an aversion that lasted for over a week. In another, cats given a meal to which the emetic lithium chloride had been added refused to eat the food that the meal consisted of, even when it was not adulterated (Mugford, 1977). This refusal persisted for several days, and subsequently only small amounts were eaten; the cats then seemed to be adopting a strategy similar to that described above for neophobia, as a result of which the aversion induced by the emetic was gradually cancelled out as the cats relearned that the food was now safe. More than one unpleasant experience following a particular food may result in that food being rejected for many months. It is not known for the cat whether such experiences can also enhance the probability of neophobia, but this seems possible.

As in the rat, such aversions extend to nutritionally inadequate diets, as well as to those containing toxins. The cat's heavy requirement for thiamine, particularly when it is on a carbohydrate-rich diet, has been noted above. The first symptom of a thiamine-deficient diet is often anorexia, apparently due to a rapidly learned dietary aversion (Everett, 1944). Diets that contain an imbalance of essential amino acids can produce a similar effect.

Learning about calories

Cats are solitary hunters, and tend only to take prey that is considerably smaller than themselves. Wolves need to be able to gorge themselves on their large communally hunted prey while it is available, and this is probably the origin of the notoriously insatiable appetite of some types of domestic dog. A solitary hunter needs to stay at an optimum weight to be successful, and so it is comparatively unusual to find an obese house cat. The majority of cats seem to defend their set-point weight quite accurately; their appetite appears to be affected within a few hours of eating a meal that contained more or less calories than expected. Some exceptions have been found to this general rule; in experiments where celluflour or kaolin were used to dilute the calorie content of foods, the cats failed to compensate, but this was probably due to an adverse effect of the diluents on palatability, to the extent that the most dilute foods were barely acceptable. Dilution with water results in accurate compensation, as does calorie enrichment with fat, so it appears that cats have accurate internal measures of the calorie contents of their foods (Fig. 6.4; see also Kane, 1989). The same may not be the case for protein content, which is unlikely to become limiting in a normal carnivorous diet. Even when the source of protein is reasonably palatable, kittens seem to be unable to select an optimum level

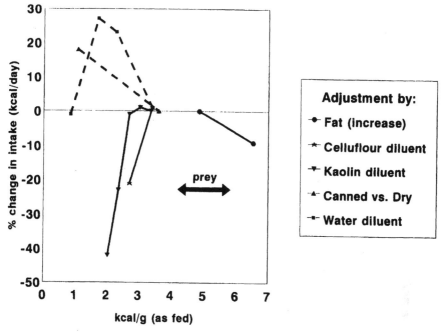

Fig. 6.4. The results of five different studies of calorie compensation by cats, expressed as the percentage change in daily calorie intake compared to the standard or control treatment. Two of the studies used water as the main diluent (dashed lines); three used solid materials to dilute or enrich the calorie levels (solid lines). The approximate range of calorie densities in normal prey is indicated by the arrow. (Redrawn from Bradshaw, 1991.)

of protein from several choices (Cook *et al.*, 1985); by contrast, rat pups are highly selective in the same situation. The physiological mechanisms underlying nutrient selection in general, and calorie regulation in particular, have scarcely been investigated in the cat, but in common with other mammals the hypothalamus plays a major role in controlling appetite. The metabolic cues that trigger changes in the hypothalamus may, however, be rather more specific and reflect the cat's carnivorous lifestyle. An inhibitor of glucostatic mechanisms, 2-deoxy-D-glucose, can cause short-term increases in food intake, as the increase in glucose levels in the blood that follows meals and inhibits continued food intake goes undetected. The normal food of cats is lower in carbohydrate than that of rats, and correspondingly it is found that cats are more sensitive to 2-deoxy-D-glucose than are rats (Jalowiec *et al.*, 1973). However, there is some evidence that low levels of sugar added to the diet are somehow not detected as calories by cats. Coupled with the study of protein selection mentioned above, this suggests that regulation of specific nutrients may differ considerably between the cat and omnivores such as the rat.

Hunting and Predation

Few domestic cats obtain all their food by hunting – most feral cats scavenge extensively – but nevertheless cats are efficient predators. Their hunting behaviour has come under the scrutiny of biologists for several reasons. First, on some oceanic islands, and occasionally elsewhere, cats can cause serious damage to seabird colonies, and the ecological effects of their hunting will be dealt with later in this chapter. Second, cats are often accused of being 'ecologically surplus killers', continuing to hunt when they already have a surplus of food available, and so they make ideal subjects for the study of the relationship between hunting and hunger. Finally, they are by far the most accessible of the small cats (*Felis* spp.), and so comparisons of the typical hunting behaviour of this genus with that of other Carnivora tend to be based on the domestic cat. Leyhausen's monograph (1979) covers the latter approach in considerable detail, and only a summary will be included here.

Predatory Behaviour Patterns

Once a cat has located its prey it will usually approach it rapidly in a crouching posture, the typical 'stalking run', making use of cover where available. On arriving within a few metres, the cat will often drop into a watching posture, in which its body is pressed flat to the ground, its fore legs are drawn back so that its forepaws are beneath its shoulders. The head is stretched forward and the ears are held erect and pointing forward. This may be followed by a second stalking run, presumably because the cat judges that the prey is not sufficiently close for an effective pounce. Once within striking distance of the prey, the cat prepares to spring, chiefly by moving its hindpaws further back. While in this position, the hind feet may be raised and lowered alternately, a movement seen in a much more exaggerated form in the mock pounce of kittens, when it can cause the whole of the hindquarters to wobble violently from side to side. The smaller movements of the adult, and the tail twitching that often also occurs at this stage, appear to be maladaptive in that either could alert the intended prey to the presence of the predator. However, since both sources of movement are at the rear of the cat, they may be screened from a targeted prey at ground level by the forequarters of the cat, and so have no impact on its conspicuousness. It is possible that the 'treading' movements stave off fatigue in the muscles of the hind legs, fulfilling a similar function to the constant changing of the stride while galloping, described in Chapter 1. Cats watching unattainable prey, such as birds that are close by but too high up to be worth striking at, may pull back the corners of their mouths and make an irregular chattering sound with their teeth. This has been explained as a displacement activity, but since this sound is never made in any other context it does not fall within the strict definition of a displacement activity, which should be recognizably derived from behaviour patterns seen under other circumstances (McFarland, 1985), and may have some other significance.

Cats rarely pounce on their prey, the exceptions occurring in long grass or similar cover, when the forepaws are used in a feline version of the fox's typical 'mouse pounce'. The final approach to the prey is usually a brief sprint, and whenever possible the final spring is kept short so that both hind feet are on the ground when the front paws strike the prey. If the prey moves suddenly at this point the spring may have to be corrected, and if this results in one or both hind legs leaving the ground at the moment of impact, the strike is often clumsy, possibly giving the prey the chance to escape. During the strike the whiskers (mystacials) are directed forwards, presumably to give precise information on the position of the prey, and compensating for the cat's poor vision at this close range. Prey that escapes into nearby cover, such as a crevice too small for the cat to introduce its head, may be 'fished' for with an extended forepaw.

The way that the prey is killed depends on what it is. Insects are usually pounced on with both paws, as are small birds. Small rodents are usually struck with one paw on the back or shoulder, and are then immediately bitten on the nape of the neck. If the prey is larger, say a young rabbit, or if the bite does not strike home, the cat will often hold the prey in its teeth and strike at it with its forepaws. If the prey still offers resistance the cat may fall on its side and rake the body of the prey with its hindclaws while holding it with its teeth and forepaws. The killing bite is ideally delivered to the neck, aimed so that one of the canine teeth slides between two vertebrae and severs the spinal cord. These teeth are equipped with proprioceptors which probably assist in locating the precise target during the bite (see Chapter 1).

After the prey is dead there is often an interval before it is actually consumed. The cat will often carry the prey to cover to eat it. Although it would probably be more efficient if the prey were grasped sideways in the teeth, the normal method of transport is to drag the prey by the nape of its neck, the original target site and evidently still a strong stimulus. This method may result in the hindquarters of the prey getting dirty, and before eating it the cat will often shake the prey in its mouth to dislodge any dirt. This shaking is quite different, both in form and intensity, to the 'death shake' used by many canids to kill their prey. Cats will also shake their prey while eating it, to loosen skin from muscle, and muscle from bone. They prefer to eat mammalian prey from the head downwards, unless the skull is too strong to be broken, in which case they will discard the head and start at the neck or abdomen. Birds are usually plucked before eating; mouthfuls of feathers are torn out and then spat away, and periodically the papillae of the tongue are cleaned of debris by what appears to be grooming of the flanks, but is actually a reverse of the usual process. Further details of hunting behaviour and hunting strategies can be found in Turner and Meister (1988).

Senses used in hunting

A great deal of the activity of kittens is directed towards learning how to hunt (Chapter 4), and so it is not surprising that hunting technique is only very loosely based on so-called Innate Releasing Mechanisms (see McFarland, 1985). The relative importance of the senses changes as the hunting sequence progresses from location of prey through to ingestion (Table 7.1). The primary sense used in locating prey is probably hearing; cats seem to have an inbuilt curiosity for high-pitched sounds such as rustlings, scratchings and ultrasonic calls. Using the latter they may even be able, with experience, to distinguish between mice and shrews, for example. Visually guided search for prey appears to rely upon the cat's appetite for the appearance of holes and crevices, which are investigated carefully. Olfaction plays a minor role, although some cats are apparently able to follow freshly laid trails of mouse urine back to occupied burrows.

Vision is used to a much greater extent in the next phase, once a visual image has been matched to a source of sound. This image is much more powerful if it is moving, and most cats require more experience before they can recognize and attack stationary prey. The killing bite is only released by furry or feathered objects, and is initially directed by the visual image of the prey's head, further guided by signals from the vibrissae (which are more heavily relied upon in the dark). The bite itself is a set of reflexes directed by trigger points around the mouth (see Chapter 3). The front-to-rear consumption of mammalian prey is mainly directed by the pile of the fur.

Table 7.1. The major senses used in hunting, in the order in which they are used, starting with the initial detection of the prey and culminating in eating.

Sense	Typical stimuli	Activity elicited
Hearing	Rustling and scratching, high-frequency call-notes	Prey-seeking
Vision	Moving object of appropriate size	Stalking run, watching, springing
Touch (vibrissae)	Contact with prey	Grasping, killing bite
Touch (mouth)	Fur/feather-like surface	Killing bite (see Chapter 3)
Sight	Head/neck/body shape	Grasping, killing, picking up to carry around
Olfaction	?	Cutting open
Flavour/texture	(see Chapter 6)	Ingestion
Touch (sinus hairs)	Direction of hair on prey	Direction of cutting and eating

The development of predatory behaviour

Many of the influences on the kitten that determine how effective a predator a cat is have already been discussed (Chapter 4). To summarize, although it has been repeatedly suggested that play, particularly those aspects that involve predatory motor patterns, will affect hunting skills in adult life, the available evidence still points towards experiences involving real prey as being the most important determinants. This may be direct experience of prey items brought back to the nest when the kittens are young, and also observation of the mother's predatory behaviour. At that stage, a single experience of killing and eating a particular type of prey can have a profound effect on both prey preferences and hunting skills.

The effects of experience seem to be strongest when comparing adult cats' reactions to rats, which are much more formidable prey than the mice and small rodents that most cats are able to catch. While some cats become competent rat-killers, many show fear reactions towards rats throughout their lives. Cats that are never exposed to rats while young, or never have the opportunity to observe an adult cat killing a rat, are extremely unlikely to become rat-killers. However, not all of those that do get the right experiences overcome their initial defensive tendencies towards rats, which appear to be determined by some inherited factors that emerge in the second month of life. These factors appear to have more wide-ranging effects than just behaviour towards large prey, because cats that show this defensive attitude not only extend it to conspecifics, but are also unwilling to explore novel environments, and to purr during human social contact (Adamec *et al.*, 1983).

Motivational Aspects

The predatory behaviour of cats is interesting for those behaviourists whose particular concern is internal causes, both because it is reputed to be disconnected from hunger, and because it is often accompanied by what appear to be elements of play.

Hunger and hunting

There is no doubt that cats will start out on a hunting expedition immediately after consuming a meal provided by their owners. It is possible that this is driven, at least in part, by the cat's desire for variety within its diet, discussed in the previous chapter, but there is no published study of the effects of diet quality, as opposed to quantity, on the frequency of hunting. Farmers who keep cats for rodent control often provide minimal supplementary food, in the belief that a hungry cat is a more efficient hunter, and there is some evidence from field studies that this may be so, although efficient pest control requires killing but not necessarily eating. There are at least three stages at which hunger could affect hunting behaviour; the stage of looking for prey, the kill itself, and the consumption of prey. Thus a well-fed cat might not spend as much time hunting as would a hungry cat, or it might hunt less 'seriously', resulting in fewer kills, or it might kill as frequently, but consume little or none of its prey.

There is little comparative information on the amount of time spent hunting by cats with different lifestyles, but well-fed house cats may typically hunt for up to a quarter of each day, while feral (unfed) cats can spend 12 out of every 24 hours searching for food (Turner and Meister, 1988). Adamec (1976) clearly demonstrated that feeding and killing are separately controlled by introducing a live rat to cats that were already feeding; almost without exception, the cats would break off from feeding, dispatch the rat, drag it alongside the feeding bowl and resume the original meal. The rat was rarely eaten, but this was predictable because in separate preference tests it was found that rat meat was less palatable than any of the foods offered. However, the most palatable food (salmon) sometimes inhibited hunting, suggesting a degree of interaction between predatory behaviour and appetite. Such an interaction is also predicted from links that have been discovered between those areas of the lateral hypothalamus concerned with killing and eating (see Chapter 3 for further discussion of the role of the hypothalamus). However, it is clear that predatory behaviour can immediately suppress feeding when prey appears; presumably the potential benefit from obtaining a second food item outweighs the risk of having the first item stolen while the cat is occupied with capturing the second.

In another study Biben (1979) showed that, while cats will engage in predatory behaviour whether hungry or not, the tendency to kill increases with hunger (and also decreases for larger prey that is more difficult to capture). Prey is often not eaten immediately after the kill, and if a second prey item becomes available immediately, it may be killed in the proximity of an uneaten carcase. Thus hunger affects the initial stages of the hunt, and the probability of a kill, but visual and auditory stimuli arising from the prey itself can override any considerations of appetite, and initiate the predatory approach. Actual consumption of the prey depends on the palatability of any alternative foods as well as hunger *per se*.

'Play' in predation

Some people perceive cats as 'cruel' because they can be seen to toy with their prey, both when it is weakened by an initial attack, but is still alive, and after it is clearly dead. Such activity might be excusable in a kitten, which will play in almost any situation, but is often seen in adults which play in almost no other context. It is now thought that much of this type of play is a displacement activity, brought about by conflicts between the need to kill, and the fear of being injured by the prey. Biben (1979) found that the frequency of play was highest in two situations (Fig. 7.1). First, if hunger had reduced the tendency to kill the prey to a minimum, small and medium sized prey were played with (hunger=fear), although this was less evident for the largest prey she tested, young rats, which were often avoided completely if the cat had just been fed (fear $>>$ hunger). If the cat had not been fed for 48 hours, it would kill young mice immediately (hunger > fear), but would play while and after attacking adult mice or young rats (fear and hunger both high).

Biben also looked for temporal associations between individual behaviour patterns within predatory sequences, and was able to divide these patterns up into three groups. One group consisted of patterns that signified either inactivity or active avoidance (and therefore possibly fear) of the prey (Ignore, Vocalize, Groom). The second group consisted of active, prey-orientated patterns which, with the exception of Bat, did not lead to the cat touching the prey, and could be grouped as non-contact play (Herd, Spring, Crouch, Rear). The third group were pre-contact patterns which included Killbite, but also the components Toss, Carry, Mouth and Bat (a paw-strike with claws extended), some of which, Toss and Bat for example, might actually reduce the probability of a successful kill. Play was similar in

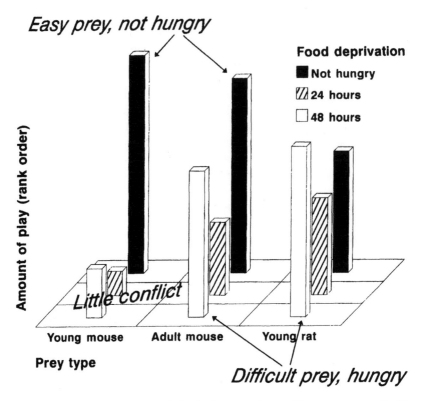

Fig. 7.1. The relative amounts of play behaviour observed in cats presented with three different types of prey, under the three conditions of food deprivation. Little play was observed when the cat was hungry and confronted with prey that was easy to catch, but when it was not hungry, or when the prey was large, play formed a higher proportion of the behaviour observed. (Data from Biben, 1979.)

quality before and after the kill; this may be because even when dead the prey still contains stimuli that release both fear and the desire to kill, and therefore still give rise to conflict. Even in the most experienced hunters, there was no relationship between the activity of the prey and the types of patterns elicited, so even these cats may be 'unsure' whether or not their victim is actually dead and therefore incapable of harming them.

Hunting Methods

Cats are opportunistic predators; they take prey in approximate proportion to its availability, and frequently supplement their diets by scavenging from human refuse and carrion. Although often said to be crepuscular predators (i.e. most active at dawn and dusk), cats are in fact very flexible in timing their hunting sorties, usually to coincide with the main activity periods of the most readily available prey. House cats appear to time their expeditions around periods when food is unavailable from their owners. Very high or very low temperatures are not favoured, so that in very hot weather cats will tend to hunt at night, and the reverse in winter.

The hunting methods of cats can be classified according to the strategy adopted, which will depend loosely upon the type of prey being sought. A sit-and-wait strategy may be adopted where there is a concentration of burrows containing suitable prey. On the other hand, small birds need to be stalked. Leyhausen considers that the domestic cat is generally a rodent specialist, because its preferred sit-and-wait strategy is much better suited to catching mice or young rabbits straying from the mouths of their burrows than it is for a bird which can move in any direction and in three dimensions. None the less, some individual cats do seem to become specialist bird hunters. Colonies of burrow-nesting birds can be raided by cats going down the burrows themselves if they are wide enough.

Cats are reasonably successful predators, since between 40% and 65% of free-ranging cats have identifiable prey in their stomachs (Turner and Meister, 1988). While the availability of prey has a large effect on what is taken, larger prey is often captured at a lower rate (measured as pounces per catch). Despite this, the larger size of rabbits means that they are actually more rewarding (measured as calories per pounce) than small rodents. An optimizing predatory cat should therefore take more rabbits than rodents (if both are equally available), and this does seem to be true of feral cats that depend on hunting. House cats, on the other hand, take fewer rabbits than optimization would predict, presumably because at their lower levels of hunger fear overcomes the urge to kill larger prey.

Turner and Meister have also been able to show that queens with kittens to feed are much more efficient predators than other cats. They followed 143 hunting expeditions made by 23 farm cats, which were provided only with bread and milk and table scraps by their owners. The mother cats caught prey on average once every 1.6 hours, while the others were only successful just over twice every 24 hours. The average duration of hunting excursions, about half an hour, was the same in both groups, but the mothers travelled twice as fast. This did not reduce the number of potential prey-containing sites that they discovered, which they therefore located at twice the rate attained by the others. They spent only about a minute investigating each location before moving on to the next, compared to about three minutes for the non-mothers, and were more successful at actual capture of prey (3.4 pounces per capture for mothers, 12.3 for non-mothers). The presence of kittens in the nest stimulates the bringing-home of prey, possibly via the hormone folliculin. When the kittens are newly weaned, the prey is killed before being transported, but this is replaced later by disabled but live prey (see Chapter 4). Cats without kittens, including intact toms, will also occasionally bring prey back to their primary 'home', but the reason for this is not known.

Ecological Aspects of Predation

Wildlife biologists are frequently suspicious of the damage that cats may do to local populations of small mammals, reptiles and small birds. There is no doubt that on species-poor oceanic islands, the introduction of feral cats can have devastating effects on ground-nesting birds that depend on the absence of predators to raise their young successfully. For this reason a considerable body of data is available on the types of prey eaten by cats, and the likely effects of this predation on prey species. Little of these data contains much behavioural information, and will only be summarized here. A detailed account is available in Fitzgerald (1988).

Species taken

Most of the information available on the prey taken by cats actually refers to what is eaten, and omits prey items that are caught and not eaten. Exceptions are reports of prey brought home by house cats, and the uneaten remains of prey found in the field in those rare situations where cat kills can be confidently distinguished from the victims of other predators, such as on oceanic islands.

Mammalian remains are commoner than remains of birds, confirming the idea that cats are specialized predators of small mammals. This does not vary from one part of the world to another, but reptiles are rarely a significant part of the diet at latitudes above 35°. Invertebrates, including insects, spiders, isopods, crustaceans and molluscs, are often recorded, occasionally in large numbers, but their generally small size means that they are unlikely to make a major contribution to overall energy intake, except perhaps in juvenile cats that are not sufficiently mature to catch many mammals.

Among the mammals, the main species taken vary between the continents. In North America and Europe, voles are often the most numerous, particularly the common vole *Microtus arvalis* and the field vole *M. agrestis*. Young rabbits and hares are also common, and because they are larger than voles can make the major contribution towards weight of food eaten. Murid rodents, including mice *Mus musculus* and young rats *Rattus* spp., are less commonly eaten than voles, possibly because they are less palatable, but house cats frequently bring these back from the field uneaten, so they are probably often killed but not eaten. Certainly many farm cats would not earn their keep if they restricted their hunting to rabbits and voles. Shrews (species of *Sorex* and *Talpa*) are certainly unpalatable and are rarely eaten. In North America, but not in Europe, members of the squirrel family (Sciuridae) are also common prey, including ground squirrels and chipmunks. In Australasia introduced rabbits, rats and mice are supplemented by marsupials and other native mammals of similar size.

The prey taken on islands depends very much on what is available. As well as the cats, mice and rats have also often been introduced, and form part of the diet, but in contrast to the diet on the mainland, birds, mainly seabirds, become much more important on islands. These can include petrels, terns, noddies and penguins, depending on latitude and the particular characteristics of the island.

Effects on prey populations

Cats probably do more damage than anywhere else where they have been introduced to oceanic islands. Perhaps the most extreme example of this is the short history of the extinct Stephens Island wren, all 15 specimens of which were brought home by the lighthouse keeper's cat in 1948. Ground-feeding and flightless species are particularly vulnerable, as are small species of seabirds that nest on the surface, and larger species of seabirds the burrows of which are large enough to be entered by cats. Local extinctions of endemic mammals and reptiles are also well documented.

On the mainland it is generally more difficult to assess the precise impact of cats because there are usually a whole range of predators present, and prey numbers can more easily recover by immigration than on islands. For example, both cats and foxes prey upon hares, and thereby compete with human hunters, but their relative impact, both compared to each other and to other mortality factors, varies considerably from one area to another. Cat predation of songbirds is noticed because it happens during the day, in contrast to mammalian kills that mainly occur at night, but it is unlikely that cats have a major impact on small bird populations, probably no greater than the effect that other mammalian predators would have if their populations were not discouraged by man's activities.

Farmers often exploit the hunting skills of cats by keeping them as controllers of mammalian pests, as well as companions. The effectiveness of this arrangement has rarely been studied scientifically. It seems likely that cats, most of which will not attack a fully grown rat, are better at culling an invading population of young rats than eradicating an existing population that includes mature individuals. Because rats are not a preferred prey of cats, it may be important for effective rat control that alternative prey is hard to come by in the summer when rat numbers are increasing. Feral cats can also be important regulators of rabbit populations, particularly in Australia and New Zealand.

Cat densities and home ranges

Cats exploit a very wide range of habitats, and their densities can range between less than one to 200 per square kilometre. Liberg and Sandell (1988) have divided these into three categories.

1. Density > 50 cats/km^2. Urban environments with rich food supplies from a variable mixture of garbage and provisioning.
2. Density 5–50 cats/km^2. Farm cats, or feral cats feeding on highly clumped resources such as seabird colonies.
3. Density < 5 cats/km^2. Rural feral cats feeding on widely dispersed prey such as rodents and rabbits.

The home ranges of intact adult males are usually much larger than those of adult females (see Fig. 8.1); if the home range was determined entirely by considerations of food supply the average ratio of 3.5 : 1 male : female home ranges would predict a 5.3 : 1 difference in weight between the sexes, whereas males are rarely more than 50% heavier than females. The home ranges of breeding males are in fact heavily influenced by social considerations, and will be considered further in the next chapter. However, it has been observed that feral males will maintain hunting territories in particularly resource-rich areas, from which they exclude other 'subordinate' males (Corbett, 1979).

Female home ranges do seem to be determined largely by the availability of food, even though they have been found to range from 0.1 ha in a Japanese fishing village, to 170 ha in the Australian bush. Food distribution also has an effect; ranges tend to be larger than predicted when food is clumped. Farm cats that are well provisioned at home also have larger home ranges than predicted, probably because they have to travel between their resting sites in the farm and their hunting grounds. The home ranges of well-provisioned solitary house cats have scarcely been studied, but by radio-tracking a neutered female and a neutered male (which presumably did not have the home-range requirements of a breeding male), I estimated their home-range sizes to be 0.45 and 0.27 ha respectively (see Fig. 9.1). The numbers of male and female cats in the area (see Chapter 9) allowed for exclusive territories within the sexes, as suggested by Tabor (1983), so long as male and female ranges overlapped. Further spatial overlap, while exclusive territories are maintained, may also be achieved by time sharing (the communicative aspects of which are discussed in Chapter 5). However, there have been few unequivocal demonstrations of exclusive, non-overlapping territories in female cats, and the rule seems to be that if home ranges are determined by food sources, they generally overlap.

Considerable overlap of home ranges certainly does occur when females live together in a group. The social structure of these groups will be described in detail in the following chapter, but the primary reason for their formation is the existence of a large and predictable food supply, which it would not pay any one individual to defend. This is because domestic cats rarely if ever hunt in groups, and if they do (e.g. as observed of a brother and sister by Turner and Meister) there is no evidence that this enables them to take prey any larger than they would normally hunt when solitary. The food source is almost invariably of human origin, including garbage dumps, concentrations of rodents on farms, or direct provisioning (and occasionally two or all three at once). There is little interchange of females between these groups, so it is usually assumed that each group defends its resources against intruders from other groups. The home ranges of such females overlap considerably, particularly around the major food source, although each may have a non-overlapping hunting ground; for example, in one group studied by Turner and Meister the home ranges overlapped by an average 55%. The significance of group living is not restricted to range and food sharing, as will be described in the next chapter.

8

Social Behaviour

As recently as ten years ago it was commonplace for authors to assume that the domestic cat was an essentially solitary creature, that only tolerated the close proximity of its conspecifics for mating and while rearing offspring. It has often been said that the lion is the only social felid, but while it is true that the social system of the lion is complex, involving communal hunting and co-operation within both male and female groups, liaisons between members of otherwise solitary felid species have been observed. For example, cheetahs are often distributed at very low densities, and therefore opportunities for social interaction are few and far between. However, even under these circumstances small groups of male cheetahs can defend a communal territory, while other groups adopt a more nomadic way of life. Other males, and almost all females, tend to be solitary. So little is known about the behaviour of some species in the genus *Felis* that it is quite feasible that some of them are also social; for example, unexplained congregations of jungle cats *F. chaus* have been observed. It is perhaps surprising, given their familiarity, that the social structure of groups of domestic cats has only been described recently, but it is now abundantly clear that such groups are widespread, and are not artefacts of the conditions under which house cats are kept; in fact, social structure is most clearly present in groups that are barely tolerant of human company. Looking objectively at the vast range of population densities recorded for domestic cats (Fig. 8.1), it seems very unlikely that a single system of intraspecific interactions could be effective, when individual cats can find themselves spaced anything between tens and thousands of metres apart. In common with other members of the Carnivora that can adapt to a wide range of population densities, the particular social system that prevails among each population of domestic cats varies according to the ecological circumstances in which they find themselves (Macdonald, 1983). Essen-

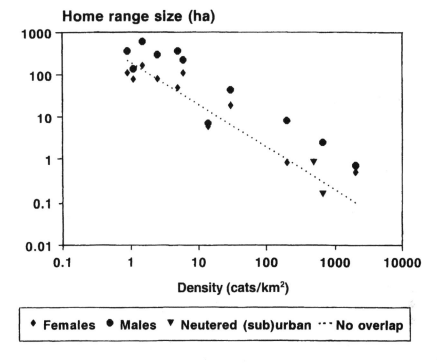

Fig. 8.1. The relationship between home range size and cat density, for entire males, females and neuters. The dotted line indicates the size of home range expected if each part of the available space was allocated to the home range of one male and one female; points well above this line indicate overlap between the ranges of members of the same sex, points well below indicate that not all the available space is used. (Data from Liberg and Sandell, 1988, with additions from Chipman, 1990, and my own data (see text).)

tially, groups may be formed when the availability and dispersion of food allows two or more individuals to live in close proximity, and on all the occasions that this has been documented much of this food has stemmed from man's activities. This inevitably raises two questions; whether such coalitions can ever occur without man's tacit collaboration, and if not, how the necessary behaviour patterns evolved, unless they are a by-product of domestication. This issue, along with many aspects of the detail of cat–cat relationships, requires further research.

Solitary Cats

The question of whether cats kept singly in households are truly solitary will be left to the following chapter, but it can be argued that, like the dog that appears to perceive its human owners as part of its 'pack', many cats act as if their human keepers are other cats. Truly solitary cats, that have little conspecific or human contact for much of the year, have not been studied in much detail, not least because they are difficult to locate and harder to approach. The European wild cat *F. s. silvestris*, although not the closest wild ancestor to the domestic cat, is thought to be almost entirely solitary, and may be genetically predisposed to be so, since even its kittens are very difficult to domesticate. Largely solitary populations of *F. s. catus* are also known, for example those in the bush of SE Australia, and on many uninhabited islands. Generally, these populations support themselves by hunting, and it is rare for prey suitable for domestic cats to be highly abundant in any one location for long enough for a social group to develop. Under these circumstances cats are rarely seen in the company of other cats, except for male/female pairs at oestrus, and females with juveniles. The adults are usually territorial to some extent, although the mechanism whereby such territories are maintained, given that individuals so rarely encounter one another, is not clear. When food is more patchily distributed, but each patch or group of patches is still insufficient to support more than one cat, home ranges can overlap quite extensively, but some system of temporal separation may then operate, so that two cats rarely hunt the same area at the same time. Scent marking has been implicated in maintaining this 'time-sharing' arrangement (see Chapter 5).

As discussed in Chapter 7, the home ranges of females encompass sufficient food and shelter for their needs and those of their offspring while they are dependent. Even when cats are well dispersed, the home ranges of males are larger than those of females (Fig. 8.1), 3.5 times larger on average. Male ranges of up to 10 square kilometres have been recorded, and it is possible that some males, labelled as 'transient' in most studies, either have even larger home ranges than this, or are more or less nomadic (Liberg and Sandell, 1988). Male ranges appear to be dictated by the availability of breeding females, whether those females are solitary or social, and the degree of competition for those females; the factors determining the size of individual male ranges will be discussed below.

Group-living Cats

Both wide-ranging surveys, and more detailed studies of small areas, have documented the existence of groups of domestic cats, but until recently these were considered to be mere feeding aggregations with no coherent structure. Such aggregations may occur, but when evidence for more complex sociality has been sought, long-term collaboration between individuals has usually been found. The conditions for the establishment of these groups almost always involve a localized concentration of food, arising deliberately or accidentally from human activities (Kerby and Macdonald, 1988). Some occur around garbage dumps, studied in locations as diverse as Portsmouth naval dockyard in the UK and a Japanese fishing village. Others are more direct products of provisioning, such as the semi-wild populations often found on industrial and hospital sites. Farm cats are provisioned both directly and indirectly, by direct handouts of food, by the concentration of rodents in grain stores, and sometimes by the theft of food intended for livestock.

Most of the groups that have been studied have turned out to consist of females, usually related, together with their offspring, including immature males. The size of the groups seems to be very variable; in a survey of 300 colonies on industrial sites in the UK, most comprised between one and ten individuals, but 7% contained over 50 cats. The critical factors determining the size of the groups are the availability of food, infant mortality due to feline panleucopenia and other viruses, and direct killing of adults by man. Spontaneous movement of females between groups seems to be rare, probably because while females within a group are generally tolerant of one another, they usually attack outsiders on sight, both males and females, and these attacks generally become more intense when there are young kittens in the group. Mature males are only loosely attached to any group, and as with solitary cats, their home ranges tend to be larger than those of females.

Sexual Behaviour of Females

Apart from the contact between mother and offspring, described in Chapter 4, the only essential component of social behaviour required of a solitary female is that leading up to mating. In solitary individuals, there will be a strong territorially based tendency to attack any cat, and one of the main functions of courtship behaviour may simply be to bring the sexes together without fighting for long enough for copulation to take place. Even within a colony of cats, where all individuals are familiar to one another, if a male shows more than a fleeting sexual interest in an anoestrous female, the female will spit and strike out with her claws. In proestrus, the behaviour of the female changes, first subtly as a tendency to move about more than usual, and then as an increase in object rubbing (see Chapter 5). Males that approach at this stage are greeted with less hostility than before, but prolonged contact is still not tolerated. Over the next 24 hours the rubbing increases in intensity, and persistent bouts of rolling occur, accompanied by purring, stretching, and rhythmic opening and closing of the claws. Males are now permitted close to the female, and may be allowed to lick her, but at this stage any attempts at mounting result in a considerable display of aggression. Complete sexual receptivity does not ensue until the beginning of oestrus, which is often indicated by an abrupt change in behaviour. The rolling of proestrus is interrupted by the female adopting the lordosis position, suddenly crouching with her head close to the ground and her hindlegs treading and partly extended. Her tail is laterally displaced, uncovering the perineum, and it is at the moment that this display first appears that an experienced male will first attempt to mount. Grasping her neck in his jaws, he begins his copulatory thrusts, while the female treads backwards with her hindlegs so that the perineum is rotated further backwards and upwards, until the male achieves intromission. At this point the female usually emits a loud piercing cry, and within a few seconds jumps away from the male, and turns on him, spitting and scratching. The female then grooms her genital region, and begins to roll vigorously. Several minutes later she will adopt the lordosis position again, and this cycle of events can be repeated, with the interval between copulations lengthening, over the next one or two days (Michael, 1961).

Multiple copulations are normally needed to trigger ovulation, and without copulation ovulation does not occur. On the one hand, it has been suggested that this induced ovulation is an adaptation to solitary living, preventing the female from ovulating wastefully, before she has been able to attract a male. On the other, the whole process of proestrus, when the female is attractive to males but not receptive, and also the need for multiple matings, could also be devices to enhance competition between several males courting one female, and may therefore be an adaptation to sociality. However, the mating system of the domestic cat still presents many puzzles, as will be discussed later when intermale behaviour has been described.

Social aspects of maternal behaviour

Females tend to stay within a single social group for much of their lives; a solitary cat may occasionally join an established group, and formerly group-living females can become solitary, but migration between groups by females seems to be rather rare (Liberg and Sandell, 1988). From this, and the difference between the amicable treatment of fellow members of the group, and aggression towards outsiders, it can be inferred that group-living females defend a communal core territory, which is likely to include their denning sites, and their major source of food. However, the size and flexibility of these territories have not been systematically measured.

The most obvious co-operation between the female members of a group is the communal denning and nursing of kittens. This is such a common feature that it is perhaps surprising that it has not been studied in great detail, particularly in terms of the benefits that it may confer. While in large groups such collaborations tend to be within mother–daughter pairs, in small groups all the adult females may nurse each other's offspring, and the litters are often pooled in communal dens. The following account, taken from David Macdonald's study of such a group at Church Farm, Devon (Macdonald *et al.*, 1987), illustrates the kinds of collaboration that can take place. There were three adult females on the farm, SM and her two daughters DO and PI. In May 1987 PI gave birth to three kittens and initially nursed them unaided; when they were 18 days old DO gave birth to five kittens in the same nest. PI assisted in this birth by helping to clean the kittens, and consuming some of the afterbirth, and was seen to bite through at least one umbilicus. Both females then nursed all eight kittens, apparently indiscriminately, until the entire litter died from feline panleucopenia. When PI and DO's mother SM produced a single kitten two weeks later, PI spent a great deal of time in the nest tending the kitten, and both sisters brought prey back to the nest to be eaten by SM or the kitten (although they also 'stole' prey brought to the kitten by SM for themselves).

The following year SM and DO set up a communal nest with an unrelated mother–daughter pair, WT and TB, after WT had abandoned her own litter, in another nest, at birth. (Communal denning with unrelated females is unusual in feral cats, and may on this occasion have been due to the colony being moved to a new farm.) All four queens nursed the nine kittens produced by SM, DO and TB, although to varying extents. A week after the last of these kittens was born, a strange male was seen to enter the nest and kill six of them by biting the backs of their skulls; three of the adult females were attracted by the commotion and drove him away. The three surviving kittens (one each belonging to SM, DO and TB) were then each moved temporarily to other nests, but within two weeks all were back in the communal nest. Once they had emerged from there the three mothers appeared to treat them indiscriminately. By this stage the fourth female's (WT) involvement was minimal, but it is impossible to tell whether this was because she recognized that none of the kittens was her own, or because, as shown by her abandonment of her own litter, she was simply indisposed to provide maternal care.

The death of the first of these combined litters illustrates one certain disadvantage of this method of maternal care, that contagious diseases are likely to be fatal to all members. Several viruses are ideally suited to transmission within a communal den, and evidence for the effect of feline panleucopenia on the advantages of sociality has been obtained from the oceanic Marion Island. The virus was introduced deliberately to this island as a measure to control the feral cat population, and not only was there a four-fold decrease in the population, the proportion of cats living in groups declined also. It has been suggested that when one female has antibodies against this or another virus, and another has none, all the kittens that they share will obtain antibodies *via* the first female's colostrum. This will benefit the second female directly, and also, indirectly, the first female, if she is closely related to the second female and therefore to that female's kittens as well as her own. This is likely because communal denning is normally only recorded within mother–daughter groups; co-operation between two lineages, as described above, seems to be unusual. However, the potential advantage of antibody sharing still remains within the realms of theory.

In a study of a much larger farm cat colony, Gillian Kerby (1987; also see Kerby and Macdonald, 1988) was able to establish that females within the core group, which had their nest sites close to the food resource, and often pooled their litters, were much more successful at rearing kittens than females that raised litters singly, away from the food resource. Whereas the central females failed to rear 15% of their litters, complete mortality in the peripheral litters was 52%. After seven years of observation, some old peripheral females had no surviving descendants. Thus whatever the costs of communal denning, the costs of solitary denning seem to be higher still. The precise mechanism behind this differential success has not been narrowed down, but is likely to be due to a combination of factors, including, in this particular case, the degree of separation of the denning site from the food source, to which the kittens have to travel after weaning, rather than a simple advantage for communal denning itself.

Male Behaviour

There is very little evidence for co-operative behaviour between male cats; although Leyhausen (1988) detected loose associations between males of equal strength that he termed 'brotherhoods', the function of such groups has never become clear, and Dards (1983) never observed any amicable behaviour between any of the mature males in Portsmouth dockyard. When two unfamiliar males meet for the first time, they may initially sniff each other, but this quickly gives way to aggression, including the arched-back posture (see Fig. 5.2), growling and yowling. If one male goes on to the defensive, indicating the other's superiority, it will tend to crouch, hiss, and strike out with its foreclaws. After repeated encounters, overt aggression is reduced, but the original winner will tend to spray urine and rub objects more frequently than the loser (de Boer, 1977).

There have been various attempts to determine whether groups of males larger than two can establish some sort of dominance hierarchy, if several live in the same locality. Unfortunately, all of these studies have concentrated on measuring overt aggression and access to resources, and none progressed to the stage of elucidating subtle behavioural cues of the sort that maintain hierarchies in other social species in the Carnivora. Winslow (1938) described the behaviour of the dominant male in one particular laboratory colony; it always fed first, and attacked and attempted to mount all newly introduced cats, whatever their sex. When this male was removed,

Table 8.1. Seasonal changes in mean range size (ha) for dominant and subordinate males in the Revinge area of Sweden in 1984. Figures in parentheses indicate the numbers of individual cats whose ranges were measured. The ranges of subordinate males are thought to increase outside the mating season because the dominant males displace them from the best food sources.

	Mating season		Non-mating season	
Dominant males	218	(4)	44	(3)
Subordinate males	10	(2)	85	(2)

From Liberg and Sandell, 1988.

another started to behave in the same way almost immediately. However, no other authors have reported this method of maintaining dominance. In smaller groups, the alpha-male tends to direct most aggression towards one other member, that is assumed to be the beta-male. The degree to which a particular male is aggressive does not, however, normally correlate with priority of access to food, at least in well-fed groups.

Territorial behaviour

As stated earlier, the home ranges of males are, on average, about 3.5 times larger than those of females living under the same ecological conditions, but this apparently simple relationship conceals a great deal of variability. Outside the breeding season, male ranges tend not to overlap a great deal, unless the population density is high, but in the breeding season their ranges expand and overlap considerably, as they try to gain access to as many receptive females as possible (Table 8.1). However, if the majority of females in an area are group-living, it may not pay males to attempt to mate with more than one group, either because other groups are too far away, or because they are too well defended. Under these circumstances male home ranges may not be any larger than those of females, and this has been found in two studies, under the very high population density of a Japanese fishing village, and the very patchy distribution of cats on Swiss farms (Liberg and Sandell, 1988).

Yearling males often stay close to their mother or maternal group, but as they grow they come under increasing attack from older males. In their second or third years, they usually disperse away from their mother's home range, whether they were born to a solitary or a group-living female. Individual males that do not emigrate appear not to become sexually mature. This will have the effect of reducing the likelihood of incest, and is a common mechanism for preventing inbreeding in many species of carnivore. After dispersing, males tend to be strongly solitary, avoiding contacts with all other cats, but most eventually come to challenge the status of the 'dominant' breeding males. If the population density is low, as when cats are based on widely spaced farms, it appears to be possible for a single breeding male to secure most, if not all, of the matings. The females may interact at a high rate with such a male when he is present, and appear to be trying to prolong the male's presence in that part of his home range that encompasses their core area (Macdonald *et al.*, 1987).

Males may also expand their ranges temporarily to approach oestrous females, and this may be exaggerated when the intervention of man reduces the number of mating opportunities. For example, in one study of an urban cat population, in a 25 ha area of Manchester, UK, Paul Chipman found that only two of the 70 resident female cats had not been spayed, but there were at least 19 entire toms in the same area. The largest male home range was 6.1 ha (this male was actively avoided by others, and appeared in this respect at least to be dominant), and the average (median) was 0.88 ha. However, when the two females came into oestrus they attracted eight and ten males respectively, most of which had travelled outside their usual ranges.

The mating system

Aggregations of males around receptive females are commonplace where cat population densities are high, and their most surprising feature is that under these circumstances the males are often less aggressive towards each other than when there is no available female in the vicinity. One such group, resident in a market square in Rome, was studied in detail by Natoli and De Vito (1991). They counted over 80 cats resident in a small area (0.26 ha), fed directly by local people and scavenging refuse from nearby market stalls. Twenty-nine mature males were regularly present, of which 19 were seen mounting at least one female. These males were occasionally aggressive towards each other, but they were invariably aggressive towards outsiders, suggesting that they held some kind of communal territory. Some of the sexually active males spent most of their time in the central area (labelled 'residents'), others ('transients') were absent for prolonged periods. Within the resident males a loose dominance hierarchy could be detected, based on the number of fights won, the number of threats given, and the rate of object rubbing. Other signals, such as the number of vocal duels engaged in, and the amount of urine spraying, did not correlate with dominance rank. The resident males tended to be higher in the dominance hierarchy than the transients. However, neither dominance status, nor the amount of time spent in the central area near the females, bore any relationship to mating success. Transients and residents copulated at similar rates, because the transients spent much less time courting females and were apparently able to 'steal' matings from the residents. Even the

Table 8.2. Indicators of reproductive success by adult males in two farm cat colonies. (In a third colony, intermediate in size between these two, the same trends were measured, but the differences were generally smaller, probably due to disruption of the social structure by culling.)

	Devon	Horspath
Number of Breeders : number of Non-Breeders	1 : 1	7 : 5
% Intromissions by Breeders	100	78
Proportion of Breeders seen to intromit	100	43
Proportion of Non-Breeders seen to intromit	0	20

Data from Kerby and Macdonald, 1988.

most dominant resident allowed other males to copulate with a female while he was courting, and lower-ranking males were able to displace him from females he was attempting to mount. Although he did secure slightly more matings in total than any other individual male, in no way could he be said to monopolize matings. There seemed to be no predetermined order of access to females, and fights that broke out within a group of males courting a single female rarely resulted in any male leaving the vicinity.

In three studies of farm cats (summarized in Kerby and Macdonald, 1988) it was generally possible to ascribe a particular pattern of behaviour to the most successful males, termed 'Breeders'. A typical Breeder spent rather little time near to the main food resource, which was where the females tended to congregate, but divided his time between several other sites; when he was at the study site he kept his distance from other cats for most of the time, but interacted at a high rate, most frequently with females. He would be more aggressive than other males, and would scent mark and mate call more often also. These other adult males, termed 'Non-Breeders', spent more time near to the food resource than did Breeders, were often close to or even in contact with another cat, but actually interacted with other cats at a lower rate. They generally responded defensively towards aggression from the Breeders, and also scent marked rather rarely. The mating success of Breeders and Non-Breeders at two farm cat colonies is shown in Table 8.2. When the number of males is low, it appears to be possible for a single male to sire all the offspring; for example, for the first three seasons of observation of the Devon group described earlier, one male achieved all the successful matings, while another occupied the same position for the last two seasons, the first having disappeared. When the numbers of males and females are higher, no single male can monopolize matings, but those males classified by their behaviour as Breeders appear more likely to sire offspring than the Non-Breeders (Table 8.2). The latter often attempted to mate inappropriate recipients, such as young females, kittens or other males, so that, while they accounted for 39% of mounting attempts at the largest farm colony that was studied (25–30 cats), they achieved only 22% of the intromissions. However, even this low rate of success is likely to produce some offspring, so that in a large colony there appears to be some merit in a strategy of staying close to the adult females and 'sneaking' matings while the Breeder males are absent.

This strategy seems to be even more successful in very large colonies, such as the Rome colony described above. The transient cats were adopting a Breeder strategy, spending little time with the core group of females, but mating successfully when present, while the resident males had refined the 'Non-Breeder' strategy to secure a major proportion of intromissions. Because there were so many residents, a dominance hierarchy was apparent, presumably with its usual function of reducing the debilitating consequences of continual aggression. As far as siring kittens among the core group of females was concerned, the strategies seem evenly balanced in such a large colony. However, the transient males also had the opportunity to sire kittens by solitary females, or females in other groups nearby, while the resident males had to restrict themselves to the females in their own group. In terms of their lifetime's reproductive success, the transient ('Breeder') strategy may still be the more successful, even at high density. However, a great deal more research is needed to elucidate this, in the course of which other alternative strategies may be discovered.

One problem emerging in the interpretation of all the studies of large colonies is that the link between the number of intromissions and the actual siring of kittens is imprecise. The new techniques of DNA fingerprinting allow the paternity of individual kittens to be determined, and studies of domestic cats incorporating the results of such techniques will doubtless be reported within the next few years. Some idea of how they might be used can be gauged from the following description of a study of lions in the Serengeti National Park (Packer *et al.*, 1991). DNA fingerprints were obtained from 193 adults and 78 cubs comprising 15 prides. By comparing the fingerprints from individuals with known kinship, it could be shown that three classes of relatedness could be distinguished: close kin with co-efficients of relatedness of 0.125 or greater (the human equivalent would be first cousins or closer); distant kin with relatedness between 0.02 and 0.06 (second cousins or similar); and non-relatives. The females within a pride were always closely related. Some prides were also closely related, because not all females stay in their natal group for life, some emigrating to form new prides. By the time the founding members of each new pride are dead (maximum female lifespan is 17 years), that pride is as distantly related to its pride of origin as it is from any other. This is due to the genetic contribution of the coalitions of males that usually take over a group of unrelated females within a pride for long enough to father one or two generations of cubs. These male coalitions vary considerably in size (typically between two and six members in the Serengeti), and also in the degree to which they are related to each other. The smaller groups can contain unrelated individuals, because each is likely to father at least some of the cubs resulting from that group taking over a group of females. Within the larger male groups, DNA paternity analysis reveals that not all of the males father offspring, some effectively taking on the role of non-reproductive helpers, increasing the reproductive success of their companion males by augmenting the size of the coalition. The non-reproductive males only benefit from this strategy if they are closely related to the reproductive males, and hence the larger coalitions only contain related individuals. Each male attempts, and usually succeeds, in monopolizing a female when she is in oestrus, and DNA fingerprinting has confirmed that when there are two or more cubs in a litter, they almost invariably have the same father. Even though the females in a pride tend to come into oestrus simultaneously, there are often more males than females in the pride, so that one or more males are prevented from breeding.

Infanticide

When one coalition of male lions succeeds in ousting another group of males from a group of females, they usually kill all the cubs in the pride, thereby bringing the lionesses into oestrus more quickly than if they had completed lactation through natural weaning. Examples of infanticide by male domestic cats have been reported – one, from Macdonald *et al.* (1987), is described above. It is unclear just how common this phenomenon is, although it may be one factor causing the aggression shown by nursing females towards strange males. Since the birth interval in the domestic cat is only 4 months, whereas it is 19 months in the lion, the advantage to be gained from infanticide in bringing a female into breeding condition may not be great (Natoli, 1990). In temperate climates, the second, autumn, litter rarely produces surviving offspring, but it is possible that its chances of success are increased by bringing the time of conception forward by even a few weeks. Infanticide might also be a successful strategy when cat density is low, and adult females, food or suitable nest sites are at a premium. A non-breeding male could enhance the chances of his own future offspring surviving by attempting to kill all the offspring of the current dominant male, so that when the male lost his position, and he took over, his own kittens might be born in the best nest sites (as selected by the surviving females) and be fed sufficiently well to reach adulthood. However, it is still unclear whether infanticide is a common strategy among feral cats, or an aberration only practised by a very few.

Social Communication

We have seen that, where there is a sufficient concentration of food, cats form more or less stable groups, the basis of which is usually the co-operative rearing of kittens by related females. The role of males in these groups is still poorly understood, and may vary with population density. Despite the emphasis that has been given to studies of breeding success, the membership of the groups is generally stable outside as well as during the breeding season, and as for other social Carnivora, the social structure is maintained by a 'glue' of interactive behaviour patterns. In other species, such as the wolf, these patterns are usually related to some kind of hierarchical organization, and an individual's position in the hierarchy can usually be judged quite easily by the behaviour patterns it exhibits towards other members of its group. Some of the most diagnostic of these patterns are those indicating submission, but no such pattern has been identified in the cat, which tends to ward off aggressive approaches with defensive, rather than submissive, behaviour. Whereas a dominance hierarchy

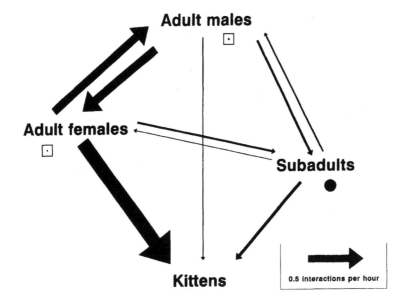

Fig. 8.2. The hourly rates of interaction between representatives of each age/sex class in a breeding colony of cats at Horspath, Oxfordshire. The relative widths of the arrows indicate the average number of interactions per hour observed to occur between average members of each age/sex class; the radii of the circles indicate the average numbers of interactions within an age/sex class, on the same scale as the arrows (small circles are shown inside squares, for clarity). Interactions initiated by kittens are excluded. (Redrawn from Kerby and Macdonald, 1988.)

between the males can sometimes be detected (based on patterns of aggression), the collaboration between females is more co-operative than hierarchical, because it is common for all the females in a group to breed simultaneously. Perhaps because of this lack of hierarchy, it is still uncertain which behaviour patterns serve to strengthen the bonds between individuals, and to build up a group identity. Among the candidates that have been proposed to fill these roles are scent marking, mutual grooming (allogrooming) and mutual rubbing (allorubbing), and these will be discussed in turn. To put these into a social context, the flow of interactions within social groups will be described first.

There is only a handful of published studies where not only has the behaviour of particular individuals been recorded, but also the partners they chose to interact with, whether those individuals were considered separately, or as representatives of a particular age/sex class. In colonies that obtain some of their food by hunting, the cats tend to visit their core area independently, neither avoiding one another nor tending to be always present with a particular partner. However, when they are present they tend to single out particular partners for interaction. The partners that each cat prefers to interact with are often those that it chooses to rest beside or in contact with, indicating that these interactions are generally affiliative rather than agonistic; the highest rate of obvious aggression occurs towards outsiders. In the Devon colony described in the section on social aspects of maternal behaviour, the three breeding females sought to position themselves close to the adult male. Within themselves, each also had preferences for the partner they rested with. In the larger Horspath colony, where there were representatives of all age/sex classes (male and female adults, juveniles and kittens), the highest rates of interaction occurred between adult males and adult females, within the juvenile group, and from adult females to kittens (Fig. 8.2). While much of the detail of the quality of social behaviour in this and other colonies remains to be elucidated, it is nevertheless evident that the interactions are highly structured, and cannot possibly be considered to support the idea that these colonies are simple aggregations around food sources.

Scent marking

The use of scent features prominently in the social lives of many mammalian species. Scents can be specific to a particular individual, are fairly stable with time, and offer the considerable advantage that they can be deposited in the environment, and later detected and decoded by a conspecific in the absence of the emitter. The scents known to be used as marks by cats include those carried in and by the urine and faeces, and those originating in skin glands on the head. Urine scent marks are known to convey individual- and group-specific information (described in more detail in Chapter 5), although the way this information is used in social interactions is uncertain.

Male cats frequently spray urine when consorting with an oestrous female, and it is possible that the rate of spraying is an indicator of mating success. Warner Passanisi, in his studies of farm cats, has found that, when attending an oestrous female, a single male sprays slightly more often than a male in a group of several males. In his farm cats, the females moved away from their core feeding area as they came into oestrus, and appeared to advertise their condition to nearby males by spray urinating, which is otherwise very unusual among females. The male always immediately overmarked the female spray with his own. Any possible role of this scent marking in the selection of sexual partners either by the male or the female remains unclear.

Allogrooming

Cats spend a great deal of time grooming, and there is no evidence to suggest that a solitary cat is any less clean than a cat that is groomed by others. The function of allogrooming is therefore likely to be primarily a social one, except in the case of young kittens that are groomed by their mother before they become competent at grooming themselves. Grooming is generally an activity that coincides with resting, and therefore it is likely that allogrooming will only be seen between cats that rest together. However, the converse, that all cats that rest together automatically groom one another, is not true, suggesting that allogrooming has some social function in addition to the general preferences for association between particular pairs of individuals. Three examples of this can be drawn from studies done by my own research group. Sarah Brown observed two neutered males in one feral colony which persistently sat together but never groomed one another. In another neutered feral group, consisting of up to 11 individuals, over three quarters of all the allogrooming was performed by one male, but this male was only involved in a third of the number of occasions that two cats sat together. Since he groomed his partners in proportion to the number of times each of these pairs sat together, his high rate of allogrooming may simply have been an example of behavioural 'style'. However, when observing a third neutered group, this one in an animal shelter, Debby Smith recorded that three adult females originally from the same litter, and an unrelated male, all regularly sat together. Within this group, 68% of the allogrooming was performed by the females to the male, and only 29% between the females. The social role of this behaviour pattern still remains something of a mystery.

Allorubbing

There is increasing evidence that one of the key behaviour patterns that cements a cat group together is mutual rubbing, in which pairs of cats rub their foreheads, cheeks, flanks and sometimes tails together. There may be two communicative aspects of this behaviour, one being the tactile signals exchanged, and the other the potential mixing of the two cats' individual scents. So far neither aspect has been investigated, so that for the present the function of this display has to be deducted from the status of the animals that initiate it, and the behaviour patterns that precede and follow it in the course of a social interaction.

In his large farm cat colony at Barley Park Farm, Oxfordshire, Warner Passanisi has confirmed an earlier discovery, that females initiate rubbing more than males do, and young animals initiate more rubs than older ones, with the net result that none of his adult males ever initiated a rub. Dividing his population into five groups, based on age and sex, a hierarchy

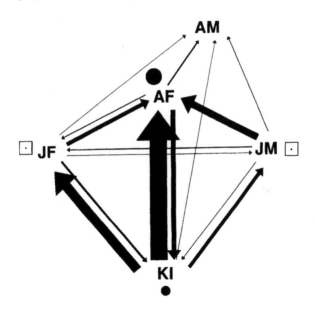

Fig. 8.3. The flow of allorubbing between the age/sex classes of farm cats in a breeding colony at Barley Park Farm, Oxfordshire. The proportions of interactions involving rubbing are indicated by the widths of the arrows (between age/sex classes) and the radii of the circles (within age/sex classes). Small circles are shown inside squares, for clarity. The proportions are not corrected for the numbers of individual cats in each age/sex class; average numbers are shown in brackets. AM = adult males (13), AF = adult females (33), JM = juvenile males (six), JF = juvenile females (nine), KI = kittens of both sexes (12). (From data collected by Warner Passanisi and David Macdonald.)

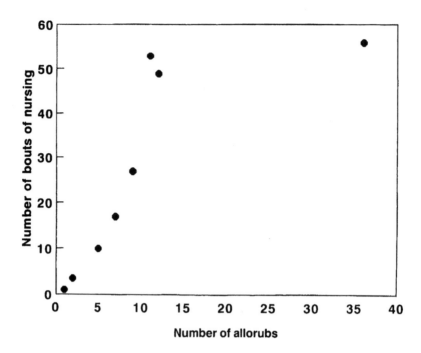

Fig. 8.4. Graph of the number of bouts of nursing versus the number of rubs, in nine adult–kitten relationships (four females and three kittens from a pooled litter; three of the possible relationships involved neither rubbing nor nursing). The point at the top right, indicating a larger than expected amount of rubbing, was between the female that nursed the most, and her single surviving kitten, suggesting that she may have discriminated this kitten from the other two, which had been born to two of the other three mothers. (From Macdonald *et al.*, 1987.)

of rubbing can be clearly seen to progress from kittens, which rub frequently on all cats except adult males, and juveniles, which rub on adult females, to adult females, which rub on each other and, occasionally, on adult males (Fig. 8.3). Another farm cat study found a higher rate of rubbing by adult females on to the single tom than between the females (Macdonald *et al.*, 1987), so variations on the basic pattern described seem to be possible.

So far as kittens in one group at least are concerned, rubbing indicates the strength of their relationship towards the lactating females that suckle them, because there is a close relationship between the number of times that each kitten rubs on a particular female, and the number of nursing bouts it receives from that female (Fig. 8.4). Which action is the cause, and which the effect, is unclear; i.e. does each female permit a kitten to suckle in proportion to the number of rubs she has received, or does each kitten

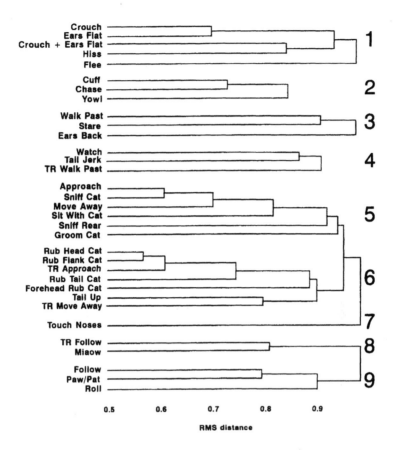

Fig. 8.5. Dendrogram showing which behaviour patterns are performed by one or other of the two cats in an interaction. Patterns which are likely to be performed by the same cat during the same interaction are indicated by lines that join close to the left-hand side of the dendrogram; patterns or groups of patterns that join close to the far right, or do not join at all, are hardly ever performed together by a given cat during a single interaction. In this way the patterns can be grouped together into clusters, most of which have self-evident functions. At the top is a defensive cluster (1), and below it a cluster of aggressive patterns (2). Below these are two clusters of patterns (3 and 4) probably rising from interactions in which one partner is unwilling to let the other approach. Cluster 5 is drawn from interactions in which a cat approaches another with its tail in the normal position, sniffs it and sits down beside it. The locomotory patterns in cluster 6 differ from those in cluster 5 by being performed with the tail raised (TR), and are often followed by allorubbing. Cluster 7 consists of a single pattern (Touch Noses) which may be highly individualistic. Cluster 8 is interaction seeking and cluster 9 is playful. (Figure produced by Sarah Brown from her unpublished data, derived from 574 interactive sequences between 11 neutered feral cats. Dendrogram produced by average linkage cluster analysis; RMS distances > 1.0 excluded.)

rub in direct response to being nursed? Whichever way round this is, rubbing does seem to have special significance as far as nursing is concerned; for example, the amounts of grooming and nursing that females give to individual kittens in a pooled litter are not closely related.

The significance of allorubbing between adults can be explored further by examining the behavioural context in which it occurs. In two separate colonies of neutered feral cats, we have found that an individual that is about to rub almost always raises its tail as it approaches the other cat (see Fig. 8.5, cluster 6). Thus, in these cats at least, the raising of the tail to the vertical seems to be one signal of an intent to rub. This is confirmed by looking at the sequences of behaviour patterns within interactions that include allorubbing (Fig. 8.6). The initiating cat almost invariably raises its tail, but the form of the rubbing itself depends upon whether the receiving cat also raises its tail. If it does, both cats usually rub simultaneously; if not, the recipient cat may either rub after the initiator has rubbed, or not at all. In a breeding colony in Portsmouth dockyard, the sequence tail-up to

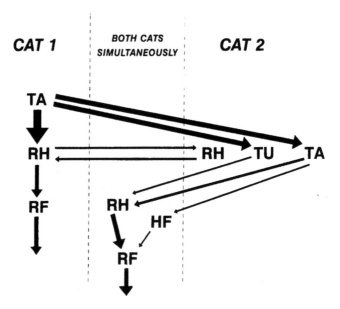

Fig. 8.6. Typical sequences of behaviour that contain head or forehead rubbing between cats. Cat 1 approaches Cat 2 with its tail raised (TA). If Cat 2 does not raise its own tail, Cat 1 rubs its head (RH) on Cat 2, which may reciprocate before Cat 1 rubs its flank (RF) on Cat 2. If Cat 2 does raise its tail (TU) or approaches Cat 1 with it tail raised, they simultaneously rub heads or foreheads (HF) together, before rubbing flanks together. Widths of arrows are proportional to frequencies (except for those emanating from RF). All transitions derived by first-order Markov chain analysis, excluding those with probabilities worse than 0.001 by Chi-square (Sarah Brown, unpublished).

allorub was often interrupted by mutual sniffing of the head (Dards, 1983), possibly because there were over 300 cats in the area, creating a need to check olfactory identity before rubbing. How these various subtleties tie into the overall flow of rubbing in a breeding colony (Fig. 8.3) has not yet been determined.

Anthropomorphically, rubbing seems to be highly affectionate, but Fig. 8.5 indicates that when cats are about to sit beside each other, they are less likely to raise their tails on approach, and are also unlikely to allorub. Thus on these occasions, when a pair of cats seem to be sufficiently confident of one another to sit together, they do not precede this with rubbing.

Rubbing seems to be used when the relationship is more one-sided, and may be the nearest the cat has to a submissive behaviour pattern that is used between individuals that are very familiar with one another. However, unlike submissive patterns in other social carnivores, rubbing is rarely seen in interactions involving any kind of overt aggression. Exceptions to this have been recorded, as when a particular tom cat was occasionally mildly aggressive towards a female that persistently attempted to rub on him (Macdonald *et al.*, 1987), but even in this case there was no evidence to suggest that the female rubbed to appease the tom.

The Functions of Domestic Cat Sociality

There has been a good deal of speculation over the advantages that cats might gain from living in groups, but to date not a great deal of evidence has been found for any of the alternatives. Given the wide range of group sizes, and the highly artificial surroundings in which some of the largest groups find themselves, it is probably not surprising that a functional explanation that appears to apply to a group in one type of situation, does not apply in another. Certainly a great deal of care is needed when extrapolating from the sociality of domestic cats to that of wild carnivores. Some types of cat groups may be more artificial than others, given that two of the three main factors regulating their size and membership are human in origin.

For females, co-operation appears to revolve around obtaining the maximum benefit for their offspring from the concentrated food-source that has allowed the group to become established. When all the females within a group are closely related, and they usually are, this can be extended to include their relatives' offspring. The best evidence for this comes from the much poorer breeding success of peripheral females compared to central females at Horspath Farm (Kerby, 1987; Kerby and Macdonald, 1988). At this farm this crucial spatial status was determined by kinship. The two central lineages had much higher breeding success than the peripheral (even though some central individuals would occasionally breed in peripheral sites), which consisted of four side-branches of the original central lineages, and two lineages based on immigrants. After several generations, these peripheral lineages would probably die out, or emigrate; meanwhile, if the central lineage bred successfully, it would tend to fragment and push some of its females to peripheral positions. In practice, this process is likely to be disturbed by man's activities (culling, the taking of kittens for pets, changes in the amount of food and shelter available), and so is unlikely to be readily apparent in all colonies.

The role of males and the mating system are still not firmly established, and these may vary considerably with the size of the colony. Small colonies tend to have single dominant males that father most or all of the kittens, and all other males in the area are reproductively suppressed either until the dominant male dies or moves away, or they emigrate themselves and gain entry into another group. In larger colonies, it seems to be difficult for a male to control the mating of several females, but it has been observed exceptionally that some males appear to form a strong pair-bond with a single female. This latter strategy is usually unsuccessful due to the low reproductive success of individual females, and in large colonies the males that father the most kittens tend to be highly mobile, and may even visit several groups of females each day. In stoats and hyenas, and other carnivores also, it is the most wide-ranging males that secure the most matings, and this possibility should not be discounted even in urban groups of cats, where such males may be extremely wary of humans, and therefore much less easy to observe than males based at a single, human-provisioned site.

The apparent lack of competition between males around an oestrous female is puzzling, but it is still not known whether or not all intromissions are equally likely to lead to offspring. It is possible that some males are able to dominate mating at the peak of the female's fertility, but allow other males to mate at other times. It is also unknown whether the order in which intromissions occur affects the probability of paternity; for example, in rats and hamsters the last male to mate sires the most young. At high population densities, females may gain in several ways from being attended by several males. They may be able to select the fittest males either by assessing their behaviour, or through sperm competition after multiple matings. Paternal uncertainty may possibly reduce the likelihood of male aggression and infanticide after the kittens are born. As a further suggestion, it might pay females living in an uncertain environment to produce as diverse a litter of kittens as possible, by mating with every available male. This could be a by-product of domestication, where man's activities may force the offspring to live in a very different environment to that occupied by their parents.

The Origins of Sociality in the Domestic Cat

Since there are no published accounts of the behaviour of group-living *F. silvestris* apart from the domestic cat, we can only speculate as to how the sociality that this species is undoubtedly capable of arose in the first place. The first question that might be asked is, if sociality is the natural state of such a familiar animal as the domestic cat, why has it only recently been recognized and characterized? Part of the answer may lie in the distortions of that social structure imposed by domestication, such that it only emerges fully in semi-dependent and dependent groups. Another part of the answer may lie in the subtlety and low frequency of cat-cat signalling; in the study of the Devon colony mentioned several times in this chapter, David Macdonald estimated that each adult female in the colony rubbed on another once every 25.3 hours. Yet another part of the answer may lie in the difficulty, when studying cat colonies, of distinguishing dominance hierarchies, which appear to be an essential component, almost a prerequisite, for social structure in other species.

As a final answer, it is undoubtedly true that many cats spend their whole lives as solitary creatures, dictated to by ecological circumstance and competition from other cats. This in turn begs a further question, concerning the ability or otherwise of individual cats, particularly females, to move from the non-social to the social state (the converse can be readily observed as large colonies outgrow their resources). The shift between social and non-social states could be achieved by at least three mechanisms.

1. All cats have the ability and behavioural repertoire necessary to live at any density within the range indicated by Fig. 8.1, and can adapt their behaviour towards conspecifics as rapidly as changes in their local cat density can take place (e.g. an individual that has been born to a solitary mother can, after living as a solitary adult itself, adapt to the conditions of living in a colony if a persistent and rich food source becomes available within its home range).

2. Adult cats can only adapt to living under social conditions if at some critical stage of their development they were actively social with more than one adult cat.

3. Sociality has its basis in inherited aspects of sociability; some individuals inherit 'sociable' characteristics, and these individuals tend to form groups, even if born to a solitary mother; others do not, and these tend to be solitary even if born into colonies.

The first possibility seems unlikely to be the whole story, based on everyday experience of house cats. The third has some support from the isolation of a component of 'behavioural style', in cats reared in near-identical conditions, relating to the equability of each individual towards other cats (Feaver *et al.*, 1986); cat 'personalities' will be discussed in more detail in the next chapter. However, there is generally very little evidence to support any of the possible mechanisms, and they are offered here as suggestions for future study.

Returning to the question of the evolution of sociality in the species *F. silvestris*, it is an unavoidable conclusion that virtually all the social groups that have been studied have relied on concentrations of food supplied by man. This raises the unlikely, but not yet disproven, possibility that sociality in the domestic cat has arisen secondarily, as a byproduct of domestication. If, for example, the original reason for domestication of the African wild cat was for rodent control in granaries, man might first have selected individuals that tolerated the close proximity of other cats, because one highly territorial cat would not have achieved the desired effect. Subsequently, those cats that displayed affiliative behaviour towards people would have been selected from the original, conspecific-tolerant, population. Those affiliative behaviours could have been derived from those shown by kittens towards their mothers, carried into the adult state by a process of progressive artificial neotenization. The process of creating this new animal might have made it even more accepting of members of its own species, so that when some individuals became feral, they were able to use this modified behaviour to form independent social groups. The apparent complexity and sophistication of cat–cat communication argues against this theory, as does the lack of correlation between sociability to man and sociability to cats in different individuals, found by Feaver *et al.* (1986). However, until such time as the origins of domestic cat sociality are found to be based on natural rather than artificial selection, it may not be wise to draw too many parallels between the functions of that sociality, and those of other felids.

The Cat–Human Relationship

The cat–human relationship appears to stand apart from any other type of human–animal interaction. The closest analogy that can be made is to the dog–human relationship, in which the dog appears to use various elements of the social behaviour of its wild ancestor, the wolf, to communicate with man (Bradshaw and Nott, 1992). The idea that will underlie the whole of this chapter is that the cat also treats its human owner as if it were a member of its social group. Unfortunately, because the social systems of the various races of *F. silvestris* are much less well understood than that of the wolf, comparisons between cat–cat and cat–human behaviour must necessarily be speculative. However, some general similarities are apparent. For example, the absence of any discernible hierarchy within animals of the same age and sex class in cats (with the exception of breeding males), and its replacement with some kind of loose co-operative structure, has its counterpart in the 'independent' relationship that cats have with their owners. Dogs appear to need to fit themselves and their human owners into a more-or-less linear hierarchy, in which each dog will normally have to occupy a subordinate position if the relationship is to be a successful one. Many cats, on the other hand, seem to have the capacity to regard their owners as equals, while others appear to treat their owners as surrogate mothers, and these are qualities that many cat owners treasure.

By living together, cats and people change one anothers' lives. Human attitudes towards cats, and the beneficial effects of cat keeping on people will not be discussed here, since this is a book about cat behaviour; more detail of this aspect can be found in Katcher and Beck (1983) and Burger (1990). In the last chapter, we have seen that the degree of sociality shown by cats is profoundly affected by food distribution, the density of con-specifics and the amount of space available for each cat. Some, and usually all, of these are different for a pet cat compared to a semi-wild feral. A pet cat living in the countryside may enjoy similar freedom of space to an independent farm cat; however, although the pet will benefit from a predictable source of food, it will be restricted in its choices of cat companions. Many town-dwelling cats have their freedom of movement restricted, either because, as is the case for many apartment-dwelling cats, they are kept indoors all their lives, or because their owners control their access to the outdoors. The welfare implications of this wide variation in the size of home ranges will be discussed in Chapter 11, but there is thought to be a higher incidence of behavioural 'problems' reported by the owners of indoor cats. On the other hand, there does not seem to be any evidence that a cat that has always been restricted to living in an apartment must necessarily behave abnormally, which is a testament to the cat's adaptability.

Cats living in semi-independent social groups are generally able to choose their partners for interaction from several individuals from each age/sex class; pet cats rarely have this degree of choice, even in multi-cat households, where the owners will usually select which cats will live together. The option to leave the household and try to establish a home base elsewhere is as available to pet cats as it is to ferals, and there is much apocryphal evidence that this does indeed happen, although no scientific study seems to have been carried out. Many cats will have little opportunity for interaction with their own kind, because it is still the norm in many countries for cats to be kept singly; while economic considerations must play a part, the myth of the solitary nature of the cat is likely to have played a part in many decisions not to keep more than one cat at a time. Again, there is no indication that cats suffer from being denied the company of members of their own species, although their situation is a highly artificial one. This is because the solitary lifestyle, which would normally only occur at very low population densities, is in this case often combined with a high density of more-or-less inimical cats. The possible effects of this on territorial behaviour will be discussed below.

Cat–Human Behaviour

Cats do not generally follow their owners about in the way that dogs do, and so interactions with their owners tend to consist of short bouts. Dennis Turner (1991) has shown that interactions that are initiated by the cat tend to be longer than those initiated by the owner, particularly when the owner is very active in starting the interactions. His data suggest that people who are keen to extract as much as possible from their relationship with a cat run the risk of actually spending less time with that cat than they would in a more casual relationship, in which the cat is allowed to make much of the running. A great deal more research is needed to find out why this might be so. In families with children, Claudia Mertens has found that while children tend to approach the family cat more than their parents do, the cat itself prefers to interact with the adults. In general, the cat–human relationship is more intense when the human partner is female. The most obvious reason for this is that it is often a female member of the household that feeds the cat. It has been shown that, while cats seem to know which member of the family is most likely to provide food, and direct their interactions towards that person when they are hungry, they are likely to be just as affectionate towards other family members at other times.

Cat–human communication

Although no statistical comparison has been carried out, it is self-evident that the behaviour patterns that cats use towards people are very similar in form to those used in cat–cat social behaviour. It is not surprising that a cat wishing to defend itself against a person will hiss and strike out as it would towards an attacking cat or, for that matter, a dog. It is probably not so obvious that a cat rubbing around its owner's legs is behaving like a juvenile cat towards an adult female (see Fig. 8.3). Some of the human-directed patterns, such as purring, and kneading with the paws, have been classified as infantile; while the latter is probably derived from the treading action with which suckling kittens stimulate the flow of milk from their mother, purring is now known to occur in a wide range of contexts even among adult cats (see Chapter 5) and must be regarded as a pattern occurring throughout life.

Cats use a wide range of other vocalizations towards humans, and some owners claim to be able to recognize a wide range of meaning and nuance in these calls (e.g. see Moelk, 1944). However, little is known about the significance of any of these calls in cat society, and hence it is difficult to put cat–human vocal communication in a biological context. Allorubbing, which appears to be both an amicable and a mildly subordinate action (see Chapter 8), is commonly directed by cats towards their owners, and it is possible that stroking is the nearest human equivalent to the reciprocation often seen in the cat–cat context. Interestingly, cats that are allowed out of doors rub on their owners more frequently than cats confined indoors, and cats in multicat households tend to rub less on their owners than single cats do (Mertens, 1991); this may be because rubbing is often used as a greeting behaviour pattern after a cat has been absent from its social group for some time. Tame specimens of *F. s. libyca* also allorub their owners (Smithers, 1968), even though they are in general much less domesticated than *F. s. catus*, so this behaviour pattern was almost certainly part of the behavioural repertoire of the earliest domestic cats.

Claudia Mertens and Dennis Turner (1988) have shown that the use of vocalizations and rubbing changes depending on whether the human responds. They allowed a cat to enter a room that contained an unfamiliar person; for the first five minutes that person was not allowed to return the cat's attempt to interact, while for the next five minutes he or she could interact without any restrictions on the form of that interaction. The level of vocalization halved in the second phase compared to the first, suggesting that the cat had initially been calling to try to induce the person to interact. In the second phase the rate of head rubbing on the person increased four-fold, while the number of flank rubs increased only slightly. However, in similar trials, but with different cats and a familiar person (herself), Sarah Brown (unpublished) found a similar increase in head rubs when she began to interact, but mainly directed at nearby objects rather than at her.

Territorial behaviour

Pet cats generally have less of a need to defend a territory than ferals, because they have a reliable source of food that is defended against other cats, if necessary, by their owners. This does not prevent neighbouring cats from entering each others' houses through the cat-flaps and stealing one anothers' food when there is no one (including the resident cat) at home. Since food is rarely a limiting resource for a house cat, it is possible that any tendency to compete for food is reduced. On the other hand, competition between pet cats for foraging space is easily observed when a new cat is introduced into an area with established cat territories. Young cats may not hold any territory as such, barring the interior of their owners' house, and may take long and circuitous routes in the spaces between the territories of other cats to reach hunting areas, such as patches of urban woodland, that contain no residents. Others may simply stay close to home, ready to avoid confrontations by retreating into their owners' houses.

The territorial behaviour of house cats, while commonly described in the popular literature, has not been scientifically studied in any detail. A survey that Sarah Lowe and I conducted of cat behaviour in a short section of a suburban street can be used to expand on the broad descriptions provided above, although it is based on a very small sample of individual cats.

The surrounding area (4.1 ha) was surveyed door to door. There was a total of 90 households, 29 in detached houses and the remainder in five apartment blocks, and 20 resident cats were found, of which two were kept almost exclusively indoors and therefore could not hold any territory out of doors. All of the cats had been neutered, and of the outdoor cats eight were female and ten were male. Three of these males, and one of the females, were resident in six houses set on either side of a section of street 80 m long that ended at the edge of a city park (Fig. 9.1). The two oldest males, one a five-year-old Persian and the other an eight-year-old short-haired crossbreed, appeared to maintain non-overlapping territories, of

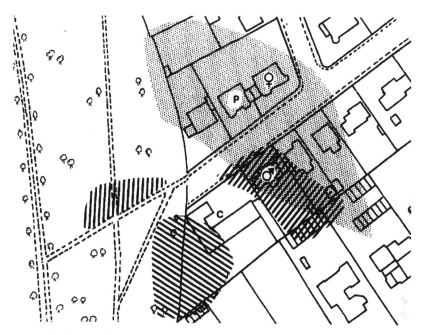

Fig. 9.1. Sketch map showing the home ranges of two house cats, one a neutered male (striped areas) and one a neutered female (stippled area). The houses that these cats lived in are denoted by the male and female symbols. Two other cats, both neutered males, lived in adjacent houses, lettered C and P. The male was never sighted in transit between its three subranges, so these have been kept separate to avoid giving a false impression of the total area that it used. (M.T.)

which the Persian's extended across the street from the north side where he was resident. The crossbreed regularly rested in a small area on the south side, immediately in front of the block of apartments where he lived, and the Persian was never sighted in this area. When the crossbreed was sighted elsewhere, he was usually in transit, so his territory may have only extended to a few tens of square metres. Interactions between these two cats were rarely observed, and when they occurred usually consisted of extended bouts of staring, so the boundaries of their territories could not be determined exactly.

Both these males would attack the third male, which was between 12 and 17 months old at the time of the study. Of all the interactions seen between the four cats (N=47), 57% were aggressive, and the great majority of these consisted of attacks by the crossbreed male, and to a lesser extent the Persian, resulting in the young male running back towards, and usually into, the house where he lived. His territory appeared to consist solely of the interior of that house. The female, which was the same age as

Table 9.1. Home ranges of cats (ha) on a 25 ha site in Manchester, UK.

	N	Median	Maximum
Entire males	17	0.88	6.10
Castrated males	52	0.076	0.99
Females	66	0.053	0.12

From Chipman, 1990.

the young male, was rarely attacked, and was the recipient of most of the non-aggressive interactions, consisting largely of play with the young male, and following, or being followed by, the Persian. Her sightings overlapped almost entirely with those of both the Persian male and the young male. This suggests that, as elsewhere, each sex tends to maintain exclusive territories, but that male and female territories can overlap completely.

The home ranges of the two young cats were measured by plotting their positions several times each day for two to three weeks; each cat was equipped with a small radio-transmitter mounted on a collar, and was initially located by triangulation of the radio signals, although wherever possible this was backed up by direct sightings, as radiotracking in built-up areas can give misleading results due to reflection of the radio signals off the sides of buildings. The young male was found to forage in two or three discontinuous areas, one adjacent to his house, and one (possibly divisible into two) in the parkland to the west (Fig. 9.1). The main discontinuity occurred around the territory of the crossbreed adult male, which he appeared to avoid by travelling quickly between the two areas through cover to the south. This results in very different estimates for his home range, depending on whether all the sightings are used (0.88 ha) or whether 95% are used (0.27 ha). The maximum home range estimate for the female was 0.85 ha, and the 95% estimate 0.45 ha; her sightings were clustered around her house, but extended mainly north-west (Fig. 9.1), including several locations where the Persian was regularly seen. Estimates of less than 1 ha for the home ranges of neutered house cats, which have no nutritional reason to hunt, are borne out by sightings of two females, which lived in separate households 50–60 m to the east, just outside, but never within, the core study area.

The home ranges of urban cats have been studied by Paul Chipman (1990) in Manchester, UK (Table 9.1). These were generally smaller than those in my own suburban study, the maximum recorded for a female (64 out of 66 were neutered) being 0.12 ha, although the overall cat density was similar (6.6 cats per hectare). The difference may partly reflect his not using radiotracking to pinpoint cats in inaccessible parts of their ranges. The 17 entire males in his 25 ha study site had ranges of up to 6.1 ha, and the largest ranges were held by males of about nine years old, with an increase from the 0.5 ha used by the young adult males, and a decrease with old age. Castrated males used ranges similar in size to those of females, except for one which had not been neutered until it was four years old; apparently its range had 'frozen' at this point.

The Socialization Process

The process by which cats become socialized to their conspecifics has been described in Chapter 4. In a domesticated animal it is essential that socialization can take place to more than one species, and the cat is no exception to this rule. Kittens of the right age will socialize towards almost any mammal that they come into close contact with, although there is some evidence that bonding to other kittens is slightly stronger than bonding to other species. For example, if a puppy is raised with a group of kittens, and each kitten is separated from its littermates so that it cries to re-establish contact, the crying is more reliably suppressed by the arrival of another of the kittens than by the puppy (Kuo, 1960). This does not seem to be due to the way that the puppy behaves in that situation, because a kitten raised with five puppies finds each one of those puppies an effective comforter. In fact, the suppression of crying in these kittens (which know only puppies) is equally as effective as the suppression of crying by a kitten in artificial litters that contain only one puppy.

The sensitive period

This socialization is most effective when the kitten is young, and so is similar to the imprinting process. This is a concept derived largely from experiments on various species of birds, such as waterfowl, in which the young become highly mobile very soon after hatching, and must therefore develop an attachment to their mother almost immediately to avoid losing touch with their family. Imprinting is the process by which they form this attachment. For example, in mallard ducks most of the species identity is built up between 12 and 18 hours after they hatch; by 20 hours all strange objects provoke a fear reaction, and the imprinting process is by then effectively over, although it can be reversed under some circumstances. Kittens develop much more slowly than ducklings, and so it is not surprising that they are able to form social attachments over a much longer period of time. This is a much less constrained process than imprinting, and so the term 'sensitive phase' is often used to describe it. Within this phase attachments are formed to any object with the right stimulus qualities (which have not been particularly well defined for the cat, but must include complex form and texture, and movement). One sensitive period occurs quite naturally in the young kitten, but even in older cats some attachments can form, or preferences for species partners change. One striking example of this is the attachment that can be formed during severe illness; there are many accounts of previously wild feral cats forming strong bonds with people that have nursed them through recovery from injury or disease. This reactivation of the socialization process is known to be induced by intense stress, mediated by the action of the hormone noradrenaline.

The precise timing of the 'normal' sensitive period in kittens has only been determined recently (Karsh and Turner, 1988). Eileen Karsh divided kittens into batches, each of which was handled for 40 minutes each day for four weeks, but starting in a different week. She measured the length of time that each cat stayed with a standard person, and found that it was highest for kittens handled between two and six weeks of age, and three and seven weeks of age. Kittens handled between the end of the first week and the end of the fifth, and between the fourth and the eighth, were less sociable. Those not handled until the seventh week were no more sociable than those that were not handled until week 14, so that it appears that the optimum period for socialization to man occurs between the second and seventh weeks, and that after seven weeks the process only occurs very slowly. Karsh did find, however, that her results were greatly affected by the 'personality' of the kitten, and that the effect of the sensitive period on the change in sociability was actually rather greater in the most timid kittens than in the others.

The amount of handling is also important; typically, a kitten handled for over one hour each day will go directly to a familiar person, climb up into his or her lap, purr, and either play or sleep. A kitten that has been handled less, for example for only 15 minutes per day, will tend to approach, head rub, and then move off. The number of different handlers that a kitten experiences can also determine its reaction to strangers. Cats that have only been handled by one person can be held for, on average, twice as long by that person than by any other, but cats with experience of four handlers will stay with any person, including a stranger, for the same length of time as the 'one-person' cats will stay with their handler. Thus there seems to be an element of generalization after several humans have been encountered, such that the one-person cat is socialized to that individual, while the multi-person cat becomes socialized to all humans that behave in broadly the same way. However, Collard (1967) found that while kittens handled by five people were more outgoing, one-person kittens were more affectionate towards their handlers, purring more often and playing for longer periods. These differences may reflect the different pathways of development that a feral kitten can go through, depending on the availability and character of adult conspecifics, which may influence its later social (cat–cat) inter-actions. However, as was discussed at the end of the previous chapter, very little is known about these processes in any context.

The character of the mother cat must not be underestimated, however. All of the studies described above were done under highly controlled conditions, with maternal influences eliminated or kept to a minimum. It is commonplace for rural queens, and some urban, to give birth in an inaccessible place and then discourage human (or indeed cat) access to the kittens until they are several weeks old. By doing this, she may well place limits on the degree to which they can be socialized to man. Her more indirect influences, which will almost certainly include the kittens imitating her reaction to people, have received little study, but are probably an important factor determining each kitten's future relationships with the human race.

Cat 'Personalities'

Every cat is an individual. This is self-evident to owners, but biologists have until recently tended to regard individual variations in behaviour, however consistently they occurred, to be a source of 'noise' to be overcome when measuring the factors determining aspects of behaviour, or the consequences that arise. This neglect probably arose because traditionally biologists have thought of boundaries between species as being much more important than boundaries between individual animals or types of animal within a species. Now that it has been re-emphasized that natural selection acts upon individual animals, or even their individual genes, and not on species as a whole, the way has been cleared for studies of the causes and functions of behavioural differences between individuals.

Having said this, the consequences of individual differences, in terms of their contribution to a lifetime's reproductive success, are extremely difficult to evaluate, and have been speculated about much more than studied directly. Mendl and Harcourt (1988) have suggested three reasons why individual differences should persist in a population. One is that two strategies might simply be equally successful, and therefore neither would confer a disadvantage. Another possibility is that each strategy is the most successful under a different set of circumstances, and if those circumstances are unpredictable in either when or where they may occur, all the strategies will tend to persist in some individuals in the whole population. Thirdly, it may pay individuals of different size to behave differently; larger animals may be able to maximize their cost/benefit strategies, whereas smaller individuals may have to 'make the best of a bad job'. The various mating strategies of feral male cats may come into the latter category. However, the domestic cat may not be an ideal model for exploring the consequences of individual variation in behaviour; just as domestication has resulted in a great diversity of coat colours, so it may also have led to greater diversity of 'personalities' than in a wild species.

Whatever the evolutionary and developmental causes of individual

Table 9.2. Three uncorrelated behavioural characteristics within a group of cats studied by Feaver *et al.* (1986). Each characteristic emerged from one or more descriptors used by human observers to rate the personalities of the cats. Several other descriptors were used (last section of table), but were less well correlated between the observers making the ratings.

---

1. Alertness

Active	Moves about frequently
Curious	Approaches and explores a change in the environment

2. Sociability

Sociable with people	Initiates proximity and/or contact with people
Fearful of people	Retreats readily from people
Hostile to people	Reacts with a threat and/or causes harm if approached by people
Tense	Shows restraint in movement and posture

3. Equability

Equable with cats	Reacts to others evenly and calmly, not easily disturbed

Not highly correlated between observers

Aggressive
Agile
Excitable
Fearful of cats
Hostile to cats
Playful
Sociable with cats
Solitary
Vocal
Voracious
Watchful

---

variations in behaviour, they presumably affect all aspects of a cat's life; for example, one aspect of the feline personality is the degree of equability towards other cats. However, it is the effect of personality on the human–cat relationship that has received the most attention, which is why this topic is being dealt with in this chapter, rather than another.

It has been shown that human observers can reliably rate cat personalities. Moreover, their ratings have counterparts in the relative frequencies of various behaviour patterns that each individual expresses under identical circumstances (Feaver *et al.*, 1986). Good correlations were found between the combined frequencies of such directly observable behaviour patterns as approaching, sniffing and rubbing on a person, and ratings of the characteristic Sociable with People (Table 9.2). This suggests that other personality measures such as Equable, which are less easily quantified from direct measures of behaviour, are nevertheless derived from genuine differences in behavioural 'style'. This method, akin to the technique of sensory profiling which is widely used in the food industry to characterize food materials and products, shows great promise for the elucidation of cat personalities.

Simpler distinctions between personalities can be made using just a single set of circumstances. For example, Mertens and Turner (1988) divided cats into three categories based upon their approach to unfamiliar persons; initiative/friendly, reserved/friendly (depending whether the cat or the human was the more likely to start the interaction) and an unfriendly type. The most significant factor affecting each cat's behaviour towards the human subjects was its personality, which was more important than the sex of the cat, or the behaviour, age or sex of the human. Turner (1991) found two types of friendly personality among Swiss house cats, one preferring play contact, and the other petting. Sarah Lowe, working in my laboratory, has recently found a similar degree of consistency with time in the personalities of 36 house cats, in the way that they behaved towards their owners in a standard situation over a period of two months; the proportions of various behaviour patterns directed towards the owners were not affected by the age or sex of the cat, nor by its origin (whether rescued or not), nor by the amount of time the cat spent out of doors. Moreover, no effect of owner attitude towards the cat could be detected, suggesting that the 'behavioural style' of house cats is relatively uninfluenced by the way that they are treated. Mendl and Harcourt (1988) have pointed out that the rating of personality is highly dependent on the method of assessment, and so it is probably too soon to make definitive statements about how many dimensions of personality cats have.

The origins of 'personalities'

Many of the differences that we can perceive between individual cats probably start during the development of the kitten, although as has been emphasized in Chapter 4, the behavioural development of the cat is a much more flexible and goal-oriented process than was once thought. Such characteristics as 'boldness' and 'nervousness' are strongly influenced by early handling; handled kittens approach novel objects much more rapidly, and spend more time near to those objects, than unhandled kittens. Moelk (1979) distinguished 'slow-quiet' and 'quick-noisy' kittens, and suggested that these characteristics would continue through to adulthood. There is some evidence to suggest that, once cat personalities are formed, they stay fairly constant for at least several years, possibly for life.

While developmental processes are undoubtedly important in determining how friendly towards people a cat will be, it has also been possible to demonstrate both maternal and paternal influences on friendliness; i.e. a friendly mother and father will tend to have friendly kittens. The maternal effect is probably due to a mixture of inherited and direct behavioural effects, but the paternal effect is likely to be entirely inherited, since male cats play little or no part in the rearing of their offspring. For example, at Cambridge University two unrelated stud males that were markedly different in friendliness were found to produce offspring which tended to be as friendly or unfriendly as their fathers (Turner *et al.*, 1986). The mechanisms that cause these differences are not yet understood, but there is no evidence in favour of the naive explanation, which is the 'gene for friendliness'.

The Effects of Neutering

There is an increasing trend for house cats to be neutered at an early age; the behaviour of entire tom cats is reasonably regarded as undesirable by the typical urban cat owner. Many owners are also unwilling to go through the trouble and expense of assisting a queen to raise kittens, as shown each year by the numbers of pregnant queens that are abandoned at cat shelters, or simply at a location far enough away from home to prevent their returning. Neutering brings about a major reduction in the behavioural repertoire, such as lack of calling by females, and the inhibition of sexual behaviour in toms that have been neutered before the onset of puberty. Despite the widespread practice of neutering in many countries, its effects on behaviour have not been studied in detail.

Apart from the inhibition of oestrus, and of course the lack of opportunities for maternal interactions, the behaviour of females does not appear to be greatly altered after neutering. The extent to which male behaviour is changed seems to depend to a large extent on the timing of castration. If this is done within the first year of life, urine spraying, aggression towards other cats and roaming, are all partially or even totally suppressed; for example, Hart and Barrett (1973) reported that 87% of males reduced their frequency of spraying dramatically after the operation. Even after a male has had aggressive and sexual experience, castration can reduce aggression, eliminate spraying, and increase the frequency of 'submissive' behaviour patterns such as rolling (de Boer, 1977). The study of home ranges done in Manchester (described above) shows that castrated males have ranges only marginally larger than those of females, and those that do range widely even after castration are likely to have been neutered after puberty. Neither spraying nor fighting are likely to be completely suppressed, but they may take place in locations away from the home base and therefore often go unnoticed. However, in multi-cat households it is often reported that one cat will spray, even if all are neutered, so there appears to be an interaction with the density of cats.

Effects on social behaviour

The practice of controlling feral cat colonies by neutering is gaining in popularity (see Chapter 10), but there has been very little study of the effects of this procedure on the social cohesion or structure of the groups. In an attempt to remedy this, Sarah Brown and I have been studying the ethology and population dynamics of neutered colonies. Apart from the obvious lack of maternal and sexual behaviour patterns, the types of inter-actions seen between colony members are comparable with those in breeding groups (Fig. 9.2). Aggression towards members of neighbouring groups seems to be reduced (Fig. 9.2), which may point to a risk that without human assistance the neutered individuals might be easily displaced by entire cats.

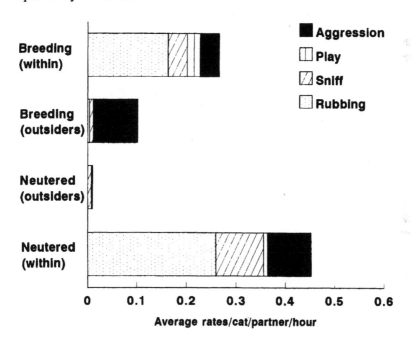

Fig. 9.2. Rates of occurrence of four behaviour patterns in two colonies of semi-independent cats, divided according to whether the recipient of each pattern was a colony member (within) or an outsider. The apparently higher overall rate of social behaviour in the neutered colony is probably an artefact of differences in the sampling methods used, but this should not have affected the relative proportions of the four patterns. (Breeding colony data from Macdonald *et al.* (1987), neutered from S.L. Brown (unpublished).)

So far as social structure is concerned, neutering does not appear to have any dramatic effects. In one free-ranging group, initially consisting of four males and seven females, we recorded the behaviour of each individual for over a year, starting from about 12 weeks after neutering. Unfortunately, there had been no opportunity to record the social structure of this group when it was breeding. The behaviour of the two adult males changed little over the period of observation; one, an orange tabby tom assumed to have been the dominant male because the majority of kittens born that year had had orange markings, initiated very few interactions throughout, but was particularly closely associated with one of the females. The other adult male was often absent from the group for several days at a time, and eventually dispersed, but was for one period also affiliative towards the same female. Otherwise his behaviour, also, changed little. Many of the changes that did take place appeared to be due to another of the females, which interacted very little with the others for the first few months after neutering, but then became increasingly aggressive towards all other members of the group. The two young males, and a third female, eventually formed an affiliative subgroup, which might not have happened if none had been neutered. The overall impression gained was that the behaviour and social position of most individuals had been 'frozen' at the time of neutering (e.g. the young males did not disperse), but that some of the effects of neutering would not have been predicted from experience of house cats.

10

Cat Welfare

The welfare of animals currently excites a great deal of public concern, and yet surprisingly little research has so far been directed towards the welfare of the domestic cat. Perhaps this is because the most obvious and pressing needs for the improvement of welfare lie among those species that are kept under the most artificial conditions, which include those of modern intensive agriculture, and those of the more old-fashioned type of zoo. The popular image of the domestic cat is that of a pampered pet, whose every whim is satisfied by its devoted, some would say excessively devoted, owner. The disturbing truth is illustrated by statistics published by Rowan and Williams (1987), showing that in the USA only a third of cats remain in the same household for the whole of their adult life, and each year a quarter of the adult cats leave their households. Some presumably move by mutual consent of the new and previous owners, but many must either migrate into the stray/feral population, or have to 'adopt' a new household because of lack of care in their previous home. Add to these the thousands of cats held by rescue organizations, and the millions of cats that lead a shadowy feral existence on the fringes of human society, and the need for more detailed examination of welfare issues becomes self-evident. Because so little research has directly addressed domestic cats, some of the issues that have been raised for other species will be discussed first, concentrating on those with an important behavioural component.

How to Define 'Welfare'

Animal welfare can be considered from a variety of standpoints, ethical and philosophical as well as scientific. There is also an important cultural

177

and personal element which influences the degree to which animals are exploited and the degree of animal suffering that is tolerated; for example in some Far Eastern cultures it is accepted that dogs can be killed and eaten, and accordingly they are often treated as production animals rather than as companions (Serpell, 1986). However, while the human toleration of poor welfare is a subjective issue, attempts have been made recently to establish a scientific basis for the measurement of welfare, and behavioural considerations are important in both establishing the causes, and measuring the effects, of poor welfare.

Welfare can be defined as the state of an individual animal as regards its attempts to cope with its environment (Broom, 1988). When conditions are difficult, individual animals use various methods to try to counteract those difficulties. Most animals have evolved mechanisms to cope with problems they are likely to encounter in their natural environment, such as extreme temperatures, desiccation, attack by predators or conspecifics, and invasion by parasites. Many of the issues of animal welfare arise because these mechanisms become either inadequate or inappropriate when the animal is faced with a man-made change to its environment. When these mechanisms are over-taxed, the animal becomes stressed and its fitness may be affected – its life expectancy or reproductive potential is reduced. The scientific definition of 'stress' is problematical, particularly when it comes to providing proof that an animal is experiencing stress under a particular set of conditions. However, it is now thought likely that all vertebrate animals experience an emotional state akin to human anxiety, because they all possess specific receptors for the anxiolytic benzodiazepine drugs (Rowan, 1988). Anxiety in wild animals probably has survival value in that it will promote attentiveness and behavioural arousal in novel situations; under man-made conditions, in which the animal has fewer options for reaction to its situation, the anxious state may be prolonged until it becomes chronic. For example, pointer dogs have been bred that show extreme anxiety when exposed to humans, even though as puppies they were reared under conditions which should have produced normal bonding to man. These dogs slink away from an approaching human, urinate and become rigidly immobile. Treatment with anxiolytic agents such as benzodiazepines eliminates these reactions, which indicates that this behavioural pathology is due to a state akin to anxiety. While in this case the stimulus was a normal one, and the abnormality resided in the responding animal, there is every reason to suppose that the reverse (i.e. a normal animal in an abnormal situation) will also lead to anxiety.

One of the problems in measuring stress is the probability that individual animals will vary in their responses. For example, some strains of hens are described as flighty because of their response to the close approach of a human, whereas other strains are termed placid because of their relative lack of response to the same stimulus. However, the supposedly placid strain exhibits a much greater and much more prolonged heart-rate response in that situation than the flighty strain does. Thus the absence of a behavioural reaction may indicate more, rather than less stress.

Assessing welfare problems

The most obvious, and least controversial, signs of poor welfare are those that arise from pain or poor health. However, an apparently healthy animal can still be suffering, due to constraints placed upon its behaviour. Suffering can arise either because the animal encounters conditions that it is motivated to avoid (aversion), or because certain conditions are absent which are necessary for a particular type of behaviour to be expressed; the animal is motivated to perform that type of behaviour but cannot, due either to physical restraint, or lack of environmental stimuli (deprivation). Deprivation is frequently seen in farm animals or wild animals placed in impoverished environments, and can result in the appearance of stereotypies, repetitive invariate sequences of behaviour of no obvious goal or function (Mason, 1991). Some of the best known of these are the 'pacing' or 'circling' behaviour patterns exhibited by wild felids in some zoos, which may arise as a result of the thwarting of hunting behaviour; despite being well fed, these animals have nothing to hunt, nor sufficient space to forage for food. However, hunting can be simulated in a variety of ways, such as by hiding food in a woodpile, or suspending it on feeding poles, to which the cats must cling to gain access to the food (Law *et al.*, 1990).

A further division can be made into short- and long-term stressors and stress responses. Short-term stress, which may cause little long-term damage, may be identified by such measures as a temporary increase in heart rate. Stressors may be ranked by comparing the responses they produce; of potential interest as far as cats are concerned are the effects that have been measured on sheep, due to being moved in a trailer (increase of 14 beats per minute) and following introduction into a strange flock (increase of 30 beats per minute), indicating that social stress can be greater than that produced by some conditions imposed by man (Broom, 1988).

The detection of long-term stress can be more problematical. Complex physiological changes can occur, including changes in the activity of the adrenal cortex, and in glucocorticoid production, but these vary both with time and from one species to another. Behavioural indicators are probably more reliable at present. Some of these are easy to observe, such as unprovoked attacks on conspecifics, and the stereotypies described above. More subtle is the 'learned helplessness' response to a restricted environment, in which the animal's threshold for response to unusual stimuli is greatly raised; the resulting appearance of placidity may be advantageous for the human handler, but is often an indicator that the animal is highly stressed.

It is thought that in many cases this condition is brought about by self-narcotization, the production of naturally occurring analgesics in the brain which dull the impact of a stressful situation (Broom, 1988). Stereotypic behaviour may also result in self-narcotization, explaining its apparently addictive quality, although there is still some doubt about this (Mason, 1991).

The alleviation of long-term stress can be approached in at least two ways. One is to ask what features of the environment are important for the expression of the whole behavioural repertoire of the animal, and then to attempt to provide these in artificial surroundings. This ethologically based approach has been particularly successful in determining the breeding requirements of pigs, and should also prove useful for the cat. The other, more psychologically based approach is to offer choices between predetermined environmental features. These choice tests, although straightforward to perform, do have their drawbacks. One is that the animals may make an unsuitable choice because at least one of the options bears no relation to the circumstances under which the species evolved. Gorging on sweet-tasting foods, seen in many mammals (though not the cat), is a case in point. Another is that simple choices may involve little effort on the part of the animal, and so give little idea of the degree of priority that the animal places on a particular resource. One way around this is to combine choices for one aspect of the environment with choices for another; for example, pigs will alter their preference for a particular type of flooring depending on whether or not they can gain access to another pig by adopting one choice or the other. Choice tests will undoubtedly have a part to play in determining the optimum conditions under which cats can be kept in shelters.

It would be helpful if we could detect pleasure, as well as the alleviation of stress, but in many animals it is even more difficult to define 'pleasure' scientifically than it is to define stress. There may be good evolutionary reasons why this should be so; an individual that communicates its state of contentment to a conspecific is inviting competition from that conspecific for the source of the contentment, whether it is a warm sleeping place or a tasty morsel of food. Purring, the behaviour pattern that immediately springs to mind as indicating pleasure in the cat, is often heard by vets from cats that are undoubtedly in pain, and so is unlikely to be a reliable indicator of good welfare.

Welfare of Domestic Cats

Cats are neither captives from the wild, nor factory-farmed production animals, so that slightly different criteria need to be applied, compared to those which are currently under consideration in zoos and in agriculture.

The wide range of degree of association with man exhibited by cats means that different welfare considerations apply to each of their lifestyles, which will therefore be dealt with separately.

House cats

Public attitudes to cats are considerably more sympathetic than they were several centuries ago when their persecution could be justified on the grounds of their association with the Devil. Nevertheless, cats are still presented at veterinary surgeries with injuries that indicate that they have been, for example, grabbed by the tail and thrown into the air. Perhaps because many cats roam freely, even in cities, many do not spend the whole of their adult lives in the same household. Statistics from the USA that support this were mentioned in the introduction to this chapter. A more detailed study of urban cats in Baltimore has indicated a considerable degree of movement between owned and stray populations; over 25% of house cats had originally been strays. Owner attitudes to sterilization are undoubtedly important; in black households in Baltimore cats were less likely to have been neutered, and were on average younger, than those in white households (Childs, 1990). This implies that sexually active cats, particularly toms, are less likely to spend their whole lives in a single household than are castrated or spayed cats.

The need for a large home range in intact males may be part of the reason for these differences, but the spatial needs of neutered cats have not been addressed directly. House cats may range over half a hectare or more if given the opportunity, depending on the density of cats in their neighbourhood (see Chapter 9). On the other hand, it is common practice for apartment-dwellers in many countries, Switzerland for example, never to allow their cats out of doors. It appears likely that the spatial requirements of domestic cats are highly flexible, and that poor welfare only results when an abrupt change occurs in the area available. This could be a reduction, as when a feral cat is trapped and confined in a shelter, or an increase, for example if an indoor cat is abandoned.

Another aspect of cat keeping that has received little attention is the possibility of a need for social contact, now that the myth of the cat's solitary nature has been debunked. Depending on the intrinsic nature of the cat–human bond, which is still not fully understood (see Chapter 9), a newly homed kitten may be able to satisfy its social needs by interaction with its owner, even if it has no contact with conspecifics. However, if such a kitten is left alone for long periods during the day, it may become distressed, in which case the most obvious solution is to home kittens in sibling pairs. This practice is already adopted by some animal charities, but as yet the scientific proof of its benefits is lacking.

Feral and stray cats

Public reaction is variable towards cats living wild, either strays that have abandoned, or been abandoned by, their owners, or ferals that have never been owned. Those who are distressed by the sight of these cats and their kittens, which often appear to be unhealthy and ill-fed, may react either by trying to provide food and care, or by demanding that they be destroyed. Other complaints relating to feral cats have been listed by Ablett (1981).

1. Fouling of gardens and communal areas of flats giving offence by sight and smell.
2. Nocturnal fighting and caterwauling.
3. Frequent finding of corpses.
4. Attacks on pets – and people.
5. Entering homes uninvited.
6. Fleas.
7. Being a 'health risk' to pets, children, and babies in prams. (Potential health hazards include ringworm, toxoplasmosis, toxocariasis and rabies.)
8. Killing and scaring of birds and ornamental fish.
9. Digging up gardens.

While some of these problems may be more apparent than real, many are genuine conflicts with the interests of people living near to feral colonies, and many have considerable behavioural components. Although the most obvious solution, and one that is often adopted, is to eradicate feral colonies by poisoning, shooting or trapping, this often has to be repeated on a regular basis because new cats move in to exploit the resource that attracted the original group. While strictly speaking the painless destruction of an animal is an ethical rather than a welfare issue, cat lovers are obviously distressed by such procedures. The 'neutering and returning' procedure, promoted in the UK by the Cat Action Trust and the Universities Federation for Animal Welfare, offers an ecologically and ethically sound alternative (Universities Federation for Animal Welfare, 1989; Passanisi and Macdonald, 1990b). As many of the cats as can be are trapped; old and incurably diseased animals are euthanized, socialized adults and young kittens are homed, and the remaining adults are neutered and returned to the original site. The removal of the tip of the left ear at the same time as the sterilization prevents the same cat from being repeatedly presented for neutering, which on balance appears to be a net improvement in welfare, although some have argued against it. Continued neutering of entire immigrants is needed, otherwise there is a risk that these cats will displace the neutered colony from the original resource, even though the cohesiveness of the original group appears to be unaffected by neutering (see Chapter 9). While the key resource may be food, Haspel and Calhoon (1990) determined that in Brooklyn the stray and feral population had access to an abundance of food, much of it provided deliberately, and that the limiting resource was shelter.

Cats in shelters

Animal shelters take in cats for several different reasons, each of which presents a different set of welfare problems. During the breeding season, a high proportion of the shelter population is likely to be kittens, either abandoned themselves or born to abandoned mothers. This is reflected in the average ages of cats from a survey of shelters in Baltimore, which at 1.3 years was considerably lower than that of the local pet population. Many people seem to have little compunction in bringing unwanted litters to shelters, the most common reason stated being 'too many pets in the home' (Rowan and Williams, 1987). Provided the staff and volunteers at the shelter have sufficient time to socialize the kittens (see Chapter 9), it should be possible to home them successfully.

Older cats can present less tractable problems. In the USA, the most common reason given for abandoning a cat bought at a pet store is some kind of behaviour problem, and while some of these may be exaggerated by their ex-owners, many cats are likely to require treatment either before or after rehoming. Other adults are brought in as strays, or are ferals taken from sites where they are no longer tolerated; many of these, particularly the latter, will be difficult or impossible to socialize to humans. Others, while approachable, may have chronic skin, respiratory or other conditions which make them unattractive as pets. Space and finance are often at a premium in shelters, and as the accommodation becomes full, euthanasia of those individuals with the worst prospects of homing may become inevitable. There are signs that the necessity for euthanasia has declined over the last 20 years. For example, between 1973 and 1982 euthanasias declined from over 20% to about 10% of all cats taken into shelters in the USA (Rowan and Williams, 1987).

Stress

When first brought into shelters, cats are often housed singly. Some obviously find this treatment, and the proximity of their human attendants, stressful, and react by extreme defensive behaviour. Others show very little behavioural reaction, and may appear superficially to be 'settled', but as mentioned above, the absence of an overt reaction may indicate more, rather than less, stress. Sandra McCune at Cambridge University has found that some of these cats not only do not groom, but often do not feed, urinate or defaecate if caged overnight, signs of acute behavioural inhibition and therefore evidence for extreme stress. Even for a cat that adapts well to a caged environment, some of the activities that inevitably occur in

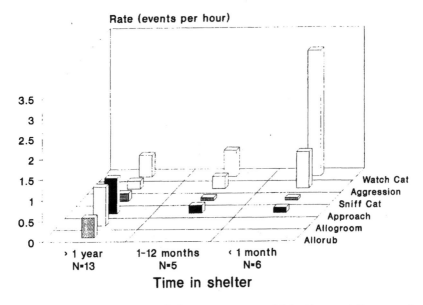

Fig. 10.1. Rates of six interactive behaviour patterns exhibited by cats (all neutered) held at a shelter in which they had access to communal outdoor pens. Averages are shown for cats that had been at the shelter for more than one year, that had had access to the communal pens for between one and 12 months, and those which had been introduced to these pens within the preceding month. Data provided by D.F.E. Smith; more information on the behaviour patterns can be found in Chapters 5 and 8.)

shelters produce measureable stress. K. Carlstead (1991, unpublished) has found that some or all of the following cause increased urinary cortisol, enhanced ACTH response (see Sparkes *et al.*, 1990), and behavioural abnormalities such as a tendency to hide: irregular feeding and handling regimen, relocation to a new environment, travel in a cat carrier, noise, and lack of petting from caregivers. A predictable environment, both in time and space, and as much time for human attention as can be spared, are both likely to improve the welfare of singly housed cats.

Socialization

For reasons of space and ease of care, adult cats may need to be housed communally rather than singly. When an individual cat is introduced into an existing group, it is likely to experience a degree of stress, which can come from three different sources:

1. residual stress from recent trapping and displacement from its original home range;

2. the restricted physical environment into which it has been placed, if it is accustomed to free-ranging;
3. the presence of strange conspecifics.

Two of my students, Debby Smith and Katy Durman, have studied the behavioural changes that take place in cats during the first days, and the subsequent weeks and years they spend in communal housing in a shelter.

Initially the cats were housed in small groups of between four and seven individuals in rooms of floor area 3.5 m², equipped with beds under and on two shelves. Newly introduced cats were aggressive towards the other occupants; on being approached they would often vocalize (hissing and growling were most common), put their ears back in the defensive position, and move away. All these patterns can reasonably be ascribed to social stress. Many of these cats would also appear to make attempts to escape, climbing and biting the bars of the doors. All of the patterns listed so far were most common in the first four days of communal confinement, and were rare in most individuals thereafter. Other behavioural measures changed more slowly, including the amount of time spent underneath the shelves, exploring the room, and sitting alert, but all had reached equilibrium by the end of the second week. By contrast to the changes in their reaction to other cats, the reaction of six of the cats, all strays, to a human handler did not alter significantly over the first month. Nor did the amount of time spent grooming, an activity which, in other species at least, is often suppressed by stress. No stereotypies were observed, and overall the relatively temporary nature of the stress-indicting behaviour patterns suggests that the long-term welfare of the cats had been only slightly reduced by their confinement.

Adult cats that remained at the same shelter for several months or years were given access to outdoor pens, area 40 m², equipped with logs, chairs and covered litter trays. Under these conditions there is a possibility that the tolerance of conspecifics observed in the rooms might give way to some kind of amicable interaction, and so this was studied in detail in a cross-section of cats that had been allowed access to these communal pens for periods ranging between a few days and seven years. Seven out of the nine cats that had been in the shelter for less than one year were never recorded as in contact with another cat, whereas nine cats, of which seven had been at the shelter for over one year, rested in contact with another cat for at least 9% of the day. Other patterns of behaviour changed also (Fig. 10.1); the most recently introduced cats were the most vigilant and aggressive, while those that had been present at the shelter for over one year were much more likely to approach other cats, and to initiate mutual rubbing and bouts of grooming. The conditions under which these affiliations had been established could not be determined, and it is possible that they had arisen during some special set of circumstances. However, these data do suggest that many cats can eventually establish sociable relationships even when confined, relationships which are likely to be beneficial in terms of their welfare. Other individuals, for reasons which are not yet clear, may never form such associations, and their welfare may therefore be best served by being housed singly.

Hand rearing

It is possible to hand rear abandoned kittens, although their chances of survival are increased the longer they are able to feed from their mothers. The potential problems of socialization of such kittens do not seem to have been studied in detail, although there are profound effects on sexual behaviour which severely reduce their chances of successful reproduction when adult (Mellen, 1988). While this is of profound importance to those interested in using the domestic cat as a model for breeding other small felids in zoos, it is unlikely to matter in a shelter where neutering before or soon after homing is the general rule. Hand raising may, however, also affect species identity, and produce an animal that is difficult to socialize to other cats; for such an individual, living with conspecifics may be actually stressful. Baerends-van Roon and Baerends (1979) have suggested that hand rearing has a damaging effect on later social behaviour even when the kitten is raised by its mother for the first seven weeks of its life. If this is so, it provides a strong argument against the homing of kittens before they are eight or nine weeks old, if they are likely to be introduced to other cats later in life.

The most economical solution to the recurrent problem of motherless kittens brought to shelters would seem to be the use of foster mothers, although not all lactating females will accept kittens that are not their own. If a kitten has to be bottle fed, attempts should be made to introduce it to kittens of the same age from the beginning of the socialization period (see Chapters 4 and 9) to ensure that it has normal relationships with other cats later in life.

Behaviour Patterns That Conflict With Domestication

Peter Neville

Association of Pet Behaviour Counsellors,
257 Royal College Street, London NW1, UK

The human–cat relationship is based on many, often contrasting factors. Indoors the cat is valued for its cleanliness, affection and playfulness, and admired for its complex play behaviour. It retains an enormous capacity to be sociable, and accepts the benefits of living in the human family and den without compromising its general self-determining and independent behaviour. Outdoors, that lack of compromise in the domestication of the cat is reflected in its ability and desire to hunt, even when well fed. The cat may view its human family as maternal figures, continuing much of its kitten behaviour into adulthood when with them (Neville, 1990, 1991). It is essential to bear in mind the strong bond between owner and cat when treating medical or behavioural problems. Most cat owners will tolerate a much higher level of disruption to their social life and household hygiene than will dog owners when pets present behaviour problems, and are far less likely to apportion blame to a cat than to a dog for its actions. Cat owners are often very sensitive to their cat's emotions and, being used to the idea that cats are not easily trained to perform set tasks, accept that their cat may not be causing problems deliberately.

Behaviour Problems in Pedigree and Crossbred Cats

A breakdown of the author's caseload referred for treatment by veterinary hospitals in London, Liverpool, Manchester, Dublin and at the Department of Veterinary Medicine at the University of Bristol Veterinary School is shown in Fig. 11.1 (Association of Pet Behaviour Counsellors, 1990). Pro rata, owners of high value breeds were more likely to seek help than owners of crossbred cats. While only 8% of British cats are of recognized

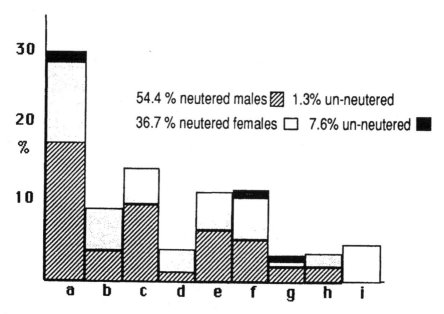

Fig. 11.1. Author's practice caseload, 1990 (Association of Pet Behaviour Counsellors, 1990).
a = Indoor spraying; b = other indoor marking (scratching, urination, middening); c = house training (loss of); d = nervous urination; e = nervous conditions, e.g. fear of visitors, agoraphobia; f = aggression to other cats; g = aggression to people; h = self-mutilation; i = other.

breeds, 44% of the caseload involved pedigree strains, 14% first-cross pedigree strains and 42% domestic short and long hairs. Of the pedigrees, 24% were Siamese, chiefly referred for problems of indoor urine spraying; 20% Burmese, chiefly for problems of aggression towards other house cats or, if allowed access to the outdoors, towards local rival cats; 13% Abyssinian, mainly for sudden breakdowns in relations between several Abyssinians sharing a home, and 13% Persians (long hairs), almost exclusively for serious house-training problems. Representatives from 17 breeds were referred. Just over half the cases (56%) were male, and only 14 unneutered cats were treated. The average age of cats presented was three years and ten months, though this is somewhat skewed by the presence of five cats over 14 years of age treated for loss of house training and excessive night-time vocalization. Most of the cats seen were between one and five years old, divided as one to two years (26%), two to three years (21%), three to four years (18%) and four to five years (11%). The most common case profile is therefore a neutered one- to two-year-old male domestic short hair or Siamese cat which lives with one other cat and which sprays or soils inappropriately indoors.

The emphasis on referral of pedigree strains may be because an owner will often have paid a relatively large sum of money for their cat and so be less willing to reject it because of a behaviour problem than if they owned a cross bred cat which would be cheaper to replace. Pedigree strains are also more likely to be housed permanently indoors, and so are more likely to present noticeable problems due to being more reactive to change within the home. The most popular of the pure breeds include the Siamese, Burmese and Persians, which are often reported by owners and breeders as being generally more 'sensitive' or emotional.

Indoor Marking (Scratching, Spraying, Middening)

Chin, head and flank rubbing are normal forms of scent marking and social communication and are encouraged by most owners from their cats (see Chapters 5, 8 and 9). Other types of marking behaviour may not be well tolerated indoors. Scratching to strop claws can usually be transferred from furniture to an acceptable sisal-wrapped post, hessian- or bark-faced board by placing this in front of the furniture and then steadily moving it to a more convenient location. Scratching as a marking behaviour is usually more widespread in the home and appears to be performed as a dominance gesture in the presence of other house cats. It should be treated in the same light as other forms of marking such as urine spraying, associative urination and defecation away from the litter tray. The latter two actions are usually performed on beds or chairs, where the owners' smells are most concentrated and from which the cat presumably perceives the benefit of associating its smell with that of a protecting influence against challenges, real or imagined. The scenario most commonly cited by owners is toileting on the bed when they go on holiday and leave friends or minders to care for their cats. Associative marking can also occur on door mats, where challenging smells may be brought in on the owner's shoes from the outside.

In contrast, urine spraying is a more normal and frequent act of marking practised by most cats outdoors, both male and female, entire and neutered. Spraying occurs from a standing position and usually a small volume of urine is directed backwards against vertical posts such as chair legs, curtains, etc. Cats usually have no need to spray indoors because their lair is already perceived as secure and requires no further endorsement. It has also been suggested that cats which spray indoors may also be more restless generally and more active nocturnally than non-spraying cats. They may also be relatively more aggressive towards the owner (Turner and Stammbach-Geering, 1990).

Of the indoor spraying cases treated by the author in 1990 (Fig. 11.1), 47% came from two-cat households; only 10% were solo house cats and

17% came from three- or four-cat households. Most were male neuters (61.5%), indicating perhaps some social inability of males especially to accept the presence of another cat in the home territory up to some variable threshold, beyond which there may be some suppressing effect on the need to spray urine. This contrasts with the previously generally held view that the greater the number of cats sharing a house, the more likely that one at least would spray. The author has visited one three-bedroom home in London used as a rescue centre for stray and unwanted cats and which houses over 140 resident pet neutered cats in addition to a continuous input of new arrivals and departures when individuals are found new homes. The owner reports that none of her cats has ever sprayed and her home does indeed smell relatively clean!

When such marking occurs indoors it is usually a sign that the cat's lair is under some challenge. The threat may be obvious in the shape of the recent arrival of a cat or dog, or a new baby, or an increased challenge from a cat outdoors, or it may result from moving or changing furniture, redecorating, family bereavement, having guests to stay, bringing outdoor objects anointed by other cats into the house, bringing in novel objects (especially plastic bags) and, most commonly of all, following the installation of a cat flap. This can totally destroy indoor security even without the obvious challenge of a rival entering the inner sanctum of the home. Protest spraying is observed particularly in some oriental breeds when frustrated or denied the attention of the owner.

Treatment

Spraying and other marking should not be considered as having a purely message function for other cats, even though feral cats are clearly able to distinguish between urine marks deposited by males, females, and neutered or entire individuals (see Chapter 5). The response of other cats to a spray mark is often to investigate it, but then to continue normally rather than run away or exhibit fear. Indoor marking may help the perpetrator feel more confident by surrounding him or herself with his or her own familiar smell. Hence cats which spray indoors may be trying to repair scent or security 'holes' in their own protective surroundings caused by change, the arrival of new objects or the addition of strange smells on new objects, cats or people, or the loss of a contributor to the communal smell that helps identify every member of the den. Spraying after cat flap installation probably occurs because the outdoors is then perceived as continuous with the indoors. The home therefore needs to be re-anointed in similar manner to the outdoors to identify occupancy and ensure that the resident encounters his or her own smell frequently. The cause should be identified if possible and the cat's exposure to any physical challenges controlled. Cat flaps should be boarded up to define den security, sprayed or middened areas should be cleaned as outlined in the 'House Soiling' section below and baited similarly with dry food. The cat should never be punished either at the time, or worse, after the event, as this furthers indoor insecurity and increases the need to mark. However, confining the cat to one room when unsupervised can create a new safe 'core' which needs no further identification by spraying and this can be expanded gradually by one cleaned, baited room at a time, initially under the supervision of the owner. Local rival cats should be chased out of the garden. Protest sprayers should be ignored and the whole relationship between owner and cat restructured so the cat only receives contact, food and affection at the owner's initiation, in a manner similar to the treatment of over-demanding or dominant dogs. This type of spraying may worsen initially before responding positively to treatment. Drug support is very much case-dependent but prescription of a tapered dose of oral progestins or sedatives for three to four weeks may assist treatment.

House Soiling

Inappropriate urination and defaecation as acts of normal elimination or as a result of nervousness, should first be distinguished from deliberate acts of marking by urine spraying described earlier or associative marking by urination and middening. Most cats instinctively tend to use loose substrate such as cat litter as their latrine when first venturing from the maternal nest, and learn by experimentation and observation of their mother that litter is a surface on and in which to excrete. Prior to this they are unable to excrete without physical stimulation from the mother. Initially this is carried out in the nest and the action enables the mother to clean all waste and prevent the kittens from soiling the nest. This process is developed when the mother carries the kittens out of the nest and licks them to stimulate excretion with the result that the majority of cats are taught early never to soil their own bed. The house is often seen as an extension of the bed and a feeding lair in adulthood. Excretion therefore normally takes place away from it, or it remains specifically targeted into a litter tray.

Poor maternal care can disrupt this latrine-association learning process and occasionally kittens are weaned without becoming house-trained, especially some Persian strains (Neville, 1990). For others, medical or emotional trauma, especially during the cat's adolescence, decreases the security of home and an initial breakdown in hygiene may then continue long after the source of the problem has disappeared or been treated. Cats that are generally nervous or incompetent may excrete repeatedly indoors rather than venture outside, and the siting and nature of the litter tray, and type of litter offered can all affect toileting behaviour. Being offered food too close to the tray will deter many cats from using it, and positioning the tray in a site that is too busy, open or otherwise vulnerable may also cause cats to seek safer places. Some cats that are normally fastidious in their personal hygiene are reluctant to use soiled or damp trays, or to share with other cats. Trays may need to be cleaned more frequently or more trays may need to be provided. Certain compressed wood pellet litters appear to be less comfortable for cats to stand on, especially for cats living permanently indoors which may have more sensitive pads to their feet than cats whose pads are toughened up by an outdoor lifestyle. Litters which release deodorizing scents when damp have also been implicated in deterring cats from urinating in the tray, possibly because of the irritation they cause when damp to the pads of the feet. Inflammation and cornification of the pads should be looked for in such cases. Litters containing chlorophyll are also reported as being unattractive to some cats and so should be changed when problems arise.

Cats may also associate pain and discomfort with their tray if suffering from cystitis, feline urological syndrome or constipation. They may then seek alternative surfaces and continue to find carpets or beds more attractive as latrines, other cats sometimes simply forget where the tray is, get 'caught short', or become lazy in old age and need perhaps more trays or easier access to them.

Treatment

It is often worth trying a finer grain commercial litter or fine sterile sand which cats, perhaps because of their desert ancestry, seem to find more attractive to use as a latrine than woodchip pellets or coarse grain litters. The position of the tray should be checked, especially relative to the position of food bowls and for security. Placing the tray in a corner or offering a covered tray, or if the cat is unwilling to be enclosed, at least a tray with sides but no roof may help to improve the security of the latrine. The tray should be cleaned less frequently to allow the smell of cat's urine to accumulate which improves identification and association and, as with dogs, the smell may stimulate animals to urinate, but the whole litter surface should not be allowed to get too dirty or too damp, as this may also deter the cat. Once per day cleansing per cat is usually adequate. For outdoor cats, up to 50% soil from the garden can be added to the litter or sand. Transfer of the use of the litter tray completely outdoors over a period of two or three weeks can be achieved by moving the tray progressively nearer the door and then out on to the step and finally into the garden.

For serious cases, confinement in a small room for a few days may help to reduce the opportunity for mistakes. Confinement in a pen where a simple choice between tray and bed can be provided may ensure that any early learning is reinforced. The cat can steadily be allowed more freedom indoors, one room at a time, when able to target excretion into the tray. Previously soiled areas in the house must be thoroughly cleaned, but never with an agent that contains ammonia as this is a constituent of urine and may endorse the idea that the cleaned area is a latrine. Many proprietary agents/cleaners may only mask the smell to the human nose and not be effective for the cat. Instead, a warm solution of a biological detergent may be followed by a wipe or scrub down with surgical spirit or other alcohol. Certain dyes in fabrics may be affected and so should be thoroughly checked first for fastness under this cleaning system. Cleaned areas should be thoroughly dry before allowing the cat supervised access. Bowls of dry cat food (with the food glued to the bottom if necessary to prevent consumption) may act as a deterrent to toileting at cleaned sites for a few days.

The cat should never be punished, even if 'caught in the act'. This makes cats more nervous and more likely to excrete in the house, even in the presence of the owner. Instead the cat should be calmly placed on its tray or outside the house and accompanied for reassurance. Timing of feeding can help to make faecal passage time more predictable in kittens and young cats and enable the cat to be put in the right place at the right time. Drug support is only usually helpful in cases of inappropriate excretion caused by nervousness. Prescription of oral sedatives for one to two weeks in addition to management may be beneficial in such cases.

Attachment/Bonding Difficulties

The critical time for socializing kittens to humans, other cats, dogs and a normal household environment, is between two and seven weeks of age (see Chapter 9). Most problems of nervousness and incompetence in adult cats would never have arisen had they been handled intensively during this period and exposed to a wide range of stimuli and experiences. Between four and 12 weeks (prior to completion of vaccination courses) they should be subjected to as complex and active a home environment as possible. Imprinting on humans in the few hours after birth probably also occurs through smell, and handling then may help produce a friendlier, tractable pet at weaning. Turner and Bateson (1988) suggest that there may be two distinct character types in cats, one with a high requirement for social contact and one for which such contact may be tolerated but not seen as an essential feature of the quality of life. The latter group seem to have a higher requirement for social play and predatory activity rather than affectionate interactions. A cat may therefore need to live with other cats and be less competent socially on its own, or need to lead a solitary life and be less able to be sociable with other cats. It is suggested that, as a result, in its relations with human owners, a cat will either have a high requirement for physical contact and petting from the owner, or will never appreciate it, even if the owner is very insistent at trying to provide it. Cats more typical of this second category may well prove less rewarding as pets, especially to those owners seeking a very affectionate relationship based on physical contact with the cat (Turner, 1991). However, these categorizations take little account of temperament changes of cats in adulthood, or resulting from personality difference between one owner and the next. Nor does it consider the fact that most cats become far more affectionate towards their owners after, for example, intensive nursing following trauma or during illness. Furthermore, improvements may also result from trying to treat individual aggressive or nervous conditions.

Under-attachment

Cats perceived by their owners to be 'under-attached' are often intolerant of the owner's proximity or approach, especially of handling, and fail to relax when held. Causes may include a lack of early socialization, over-enthusiasm on the part of the owner, trauma or necessary invasive handling during illness.

Treatment

Treatment involves increasing the bond with, and dependence on, the owners. Feeding frequent, small, attractive meals preceded by much vocal communication and encouraging the cat to follow the owner for its food is a major feature. Feeding at table level while attempting gentle handling along the cat's back only is the next step. Actions such as steadily increasing the frequency and intensity of handling, offering treats at other times and occupying favoured resting positions on the floor by the fire/radiator so that the cat comes to sit on the owner to gain access may all improve the cat's perception of the owner as rewarding. Owners should discontinue all efforts to chase the cat with a view to handling, especially if the cat seems to fit Turner and Bateson's second categorization of character, being more of a predatory and less affectionate nature generally. The more an owner tries to initiate contact with this type of cat, the less time will actually be spent in contact with the owner, according to Turner (1991). If, however, owners make themselves more attractive to their under-attached cats by offering food, titbits and toys or by lying passively in front of favourite resting places, such as the fire, and allow the cat to initiate the interaction, then the total time spent with the cat will increase. In severe cases the cat can be penned for a short time to accustom it to close human presence, and though it may seem a little bizarre, owners should try to approach the cat head first to simulate the greeting behaviour observed between friendly cats and introduce hands (perhaps otherwise viewed as threatening weapons) slowly afterwards. It is essential that owners always respond positively with affectionate touch and a calm, gentle voice to any initiating gesture the cat may make in approaching them in the home, especially for cats which are allowed outdoors. Drug support is usually not necessary except with severely traumatized cats or those unhandled before about 8 weeks, but prescription of a tapered dose of progestins may help.

Over-attachment

Over-attached cats may follow their owner constantly, perhaps crying regularly in an effort to engage them in physical contact. This is often the case in elderly cats when left alone during the night. Once the owners have responded to the cat's distress calls by getting up, perhaps because they suspect some medical problem in their pet, the cat, reassured, often simply settles back down to sleep. The over-attached cat may be agitated or nervous when isolated. Often such cats demonstrate prolonged infantile behaviour when with their owners by sucking their clothes or skin. Owners may then feel guilty about rejecting the cat's affection or fear loss of contact if they do not respond. Over-attachment is often the fault of encouragement by the owner for close association where the cat fails to lose sucking and other nursing responses after weaning. It may also occur after intensive nursing during illness or during old age with its associated increased dependence on the owners.

Treatment

Treatment involves detachment by non-punishing rejection of the cat's advances, together with periodic physical separation and replacement by alternative forms of affectionate contact for short periods of time, initiated by the owner. Provision of novel objects will help the cat learn to explore. Aversion therapy using loud startling noises, such as a jet of water, can be used in severe cases. Old cats can be offered a secure, warm bed in the owner's bedroom and will usually then remain reassured and quiet through the night without needing to cry out to gain their immediate physical attention.

Nervousness, Phobias, Separation Anxieties

Nervousness, phobias and separation anxieties are presented as a range of problems, that vary from the cat failing to adapt to 'normal' household events such as noise and visitors to lack of confidence in individual family members, failure to cope when away from the owner, and agoraphobia.

Cats may be shy and fearful if not exposed to a range of experiences and handling between the ages of two and seven weeks. This is particularly likely if they have the social play/predatory character described by Turner and Bateson and are pursued too frequently or handled too roughly at any stage. Behaviour includes becoming withdrawn and secretive, moving with a low crouching gait, reluctance to enter open space or go outdoors, inappetence or psychogenic vomiting in very severe cases. Low threshold flight reactions and defensive (fear) aggression may occur if the cat is unable to avoid the challenge. Cats with such fears may have suffered a lack of early experience or a trauma such as agoraphobia caused by fear of attacks by cats outdoors. Indeed, agoraphobia is the only recognized genuine phobia encountered in cats (Neville, 1990). Old age and its associated loss of competence may also be a factor in the development of nervous conditions.

Treatment

Systematic desensitization involving controlled exposure to known problem stimuli can be presented in low but increasing doses, while denying the opportunity to escape, so providing the possibility for habituation (Hart and Hart, 1975). With general nervousness/incompetence this is often best achieved by penning the cat indoors (or outdoors for agoraphobic cats if coupled with chasing other cats away) and forcing it to experience 'normal' household events such as the proximity of visitors, the family and other pets while protected. The cat may thus come to learn that their presence is not threatening. Frequent short meals should be offered by an increasing number of people, including visitors. Detachment from any one over-favoured member will encourage the cat to spread its loyalties to more people. Drug support is often helpful. Progestins can be administered as before, and oral sedatives during desensitization may help. Alternative medicine such as certain homoeopathic treatments and Bach's Flower Remedies may also be helpful for longer-term support of cats which are generally hyperreactive to common stimuli and nervous.

Over-Grooming and Self-Mutilation

Most cats groom their flanks or back when confused, immediately after some mild upset or when unable to avoid general threatening stimuli. The behaviour seems to have little effect on layering or quality of the coat but is a displacement activity. This is usually harmless, but occasionally a cat will over-groom in response to continued 'stress' such as the presence of too many cats in the house, acquisition of a dog, isolation from its owner, physical punishment or harassment by the owner for other behaviours such as house soiling or in response to an emotional disturbance between family members. This repetitive behaviour may have a stress-relieving function, mediated by the autorelease of opiates (see Chapter 10 and Mason, 1991). Grooming may progress to the point of breaking hair-shafts and producing a balding appearance to the flanks, the base of the tail, on the abdominal area or on the legs. In severe cases of unresolved stress or in particularly sensitive or incompetent individuals, the cat may actually pluck out large quantities of fur causing large bald patches. Often this is a secretive behaviour in its initial stages as the cat may feel more comfortable in the owner's presence and so refrain from the behaviour. In the later stages the cat may mutilate itself in their presence as well. This is an area where behaviourists and dermatologists are now conferring, as it is thought that these reactions may also be triggered as a result of flea allergy and sensitivity to diet, and occasionally from allergy to household dust, but go far past the normal groom or scratch behaviour because of some underlying 'stress'.

Actual self-mutilation of body tissue is extremely rare in the author's practice but described as a common clinical condition in Canada by Professor Donal McKeown. Such severe self-inflicted damage is usually directed at itchy infected plucked areas or, less explicitly but typically, at the tail or mouth (Luescher *et al.*, 1991). In these cases the behaviour is usually manic and occurs in frequent or occasional episodes, which may be self-reinforcing because of the euphoria engendered by the release of opiates and the preferential status of the cat compared with relaxation or facing up to an unresolvable challenge. Many cases, such as sporadic clawing at the tongue presented approximately every six months by one Burmese cat in the UK, have no obvious clinical cause. Other obsessive compulsive disorders have been described in cats, such as air licking, prolonged staring, air batting, jaw snapping, pacing, head shaking, freezing, paw shaking, and aggressive attacks at the tail or feet accompanied sometimes by vocalization (Fox, 1974; Luescher *et al.*, 1991).

Treatment

Any dermatosis should initially be investigated for medical disorders, for example, flea sensitivity, atopy or dietary allergy, and treated accordingly. Only when ruled out, or as an associative treatment, can behaviour therapy be considered. Building the competence of a cat to cope alone by restructuring relations with the owner is required if separation anxiety is suspected. Stimulation with novel objects and situations, and controlled change of husbandry patterns can also be offered. Self-mutilation can sometimes be resolved in single cats by the acquisition of another cat. The use of an Elizabethan collar for a short time may also help healing and perhaps break any learned behaviour patterns. The cat may also be distracted with sudden movement, loud noises or jets of water during severe episodes of mutilation. Generally increased levels of contact initiated by the owner may also help to define relations better and to offer more security in the home without the cat becoming over-dependent on the owner's presence.

The sedative diazepam may be given as immediate treatment to control severe episodes of self-mutilation and on a lower dose during lifestyle modification lasting several weeks. Professor McKeown reports that 20% of cats become excitable under this treatment, though this is not dose related and disappears after about five days. Anticonvulsants such as phenobarbital and mysoline, and antidepressant and anti-anxiety drugs such as amitriptyline clomipramine and fluoxetine have been employed with some success in cats in Canada, and morphine antagonists such as naloxone may inhibit the behaviour, perhaps by enabling the animal to feel the pain of its self-mutilation. Unfortunately, this drug is only active for 20 minutes or so, but longer acting versions such as Nalmefene have shown promising results under close evaluation at Ontario Veterinary College, Canada. The effectiveness of any drugs at treating such problems is believed to be influenced by the length of time the behaviour has been expressed for and the presence and ability of the owner to control conflicts and stress in the cat's lifestyle and home environment (Luescher *et al.*, 1991).

Aggression

Towards other cats

Aggression towards other cats may vary from occasional or frequent hissing or scuffling between two individuals in multi-cat households, to serious physical attack of all cats on sight, indoors or out. Despotic aggression, victimization and, most commonly, persistent intolerance of new feline arrivals to the household are all quite common. Behaviour may include physical attack, low threshold arousal in response to the sight or movement of other cats, or a total lack of initial investigatory or greeting behaviour. The cat may also be generally hyperactive and territorial. Nape biting and mounting of younger or passive cats may also be observed. Aggression rarely seems to be a defensive reaction, but occasionally attack becomes a learned policy to avoid investigation by other cats. Depending on its personality and early experiences, a cat may have an emotional need to share a home base with other cats or be more solitary. In the latter case, the cat may be able to tolerate other house cats, but never form close social ties based on mutual grooming, and resource sharing. Causes may include individual dislike or intolerance of one or more individual cats or lack of social learning or contact with other cats when young. There may be marked territorial defence reactions with failure to recognize and respond to friendly or neutral reactions of other cats, which may compound the success of early assertive or rough play with siblings. Territorial defence reactions and mutual intolerance of entire male cats, and defence of kittening areas by fertile and oestrous queens or kitten defence by mothers are normal and expected forms of aggression and not regarded as treatable. Finally, medical conditions such as hyperthyroidism, brain lesions and diet sensitivity can also cause aggression problems, but here the diagnosis and treatment clearly lies in the hands of the veterinary practitioner.

Treatment

Treatment is highly variable. Controlled frequent exposure to new arrivals by housing the original cat and the new arrival alternatively in an individual pen to allow protected introductions can be used. Distraction techniques such as bringing cats together when feeding, modification of owner relations (especially with more 'rank-conscious' oriental breeds) and instilling a hierarchy favouring the top cat in all greeting and play have all been known to help. In severe cases, rehoming may be the only safe option.

Drug support includes tapered doses of progestins, anti-androgenic injectables which may calm the aggressor (even if neutered), and progestins or sedatives and some alternative treatments may help a traumatized victim relax more during controlled introductions.

Aggression towards visitors and owners

Cats may attack people, grabbing them with claws and biting, though this is rarely accompanied by vocalization. The behaviour is often sudden and unpredictable and may be triggered by sudden movement such as passing feet or occasionally by high-pitched sounds. Defensive aggression to prevent handling is often caused by a lack of either early socialization or gentle human contact. Predatory chasing of feet and other moving body targets, territorial defence, especially in narrow or confined areas (only seen so far in oriental breeds), hyperexcitement during play, dominant aggression towards people when vulnerable (e.g. lying or sitting down), occasionally food guarding, and kitten defence against owners by nursing mothers have all been recorded. However, most problematical is the problem of redirected aggression by very territorial cats agitated by the sight through a window of rivals outdoors. Owners who approach unwittingly to pacify their cat may inadvertently stimulate an attack due to the attraction of their movement. 'Petting and biting syndrome' also occurs in many cats, but this is usually tolerated or avoided by the owner. Initially the cat accepts affection but it may then suddenly lash out, grab and bite the owner, and then leap away to effect escape. The threshold of reaction is usually high and injury slight.

Treatment

Treatment must always be considered carefully in relation to the member of the family most at risk, who are those with jerky or unpredictable movement patterns such as children and elderly relatives. Controlled exposure to habituate the cat to normal family movements and activities can help. However, therapy can be provided as stimulation, in the form of another carefully introduced cat (preferably a kitten, which may be less threatening than a more socially, territorially or sexually competitive adult), together with opportunity to go outdoors (free ranging where possible, or perhaps on a harness and lead in urban areas), frequent presentation of novel objects and concentrated play sessions (predatory chase, capture) of half an hour per day. The use of a moving target such as a ball or string to attract the release of aggression will help an excited or frustrated cat more safely, and facilitate owner intervention afterwards with less risk. Diet sensitivity and its effects on feline behaviour are little understood but may be empirically investigated by withdrawing all canned food for two weeks and replacing with fresh chicken/fish plus subsequent vitamin/mineral supplementation if matters improve. Certain modern complete dry diets may also be employed without a return of problems and with greater convenience than preparing fresh diets once any effect of diet on behaviour has been established. This treatment has been particularly valuable for those problem behaviours occurring after a change from a standard canned diet to an apparently better quality canned or 'cooked in foil' prepared diet. Access to catnip toys should be denied during treatment to preclude concomitant excitability in sensitive cats. Drug support using progestins for three to six weeks may help and some 'alternative' medicinal approaches have produced excellent responses in easily agitated oriental breeds in particular. Burmese cats seem to respond especially well to certain Bach's Flower Remedies.

Pica

Pica is the depraved ingestion of non-nutritional items. It is usually unexplained, although the eating of some house plants may be due to the desire to obtain roughage or a source of minerals and vitamins. Occasional cases are reported of cats eating rubber and electric cables, but the main problem concerns the ingestion of wool and other fabric. Wool eating was first documented in the 1950s and was thought to be limited to Siamese strains, but a recent survey of 152 fabric-eating cats by Neville and Bradshaw (1991) shows the behaviour to be more widespread. Responses to the survey showed that fabric eating was presented most by some Siamese (55% of responders) and Burmese (28%) cats, occasionally by other oriental strains and, more rarely, by crossbred cats (11%). Males are as likely as females to present with the problem and the majority of responders to the survey of both sexes were neutered. The typical age of onset of fabric eating is two to eight months. Most cats (93% in the survey) start by consuming wool, but later transfer to other fabrics. Sixty-four per cent of responders also ate cotton and 54% consumed synthetic fabric.

While some fabric eaters chew or eat material on a regular basis, others only do so in sporadic bursts. Many consume large amounts of material such as woollen jumpers and cotton towels, underwear, furniture covers, etc. without apparent harm, although surgery is required in a few cases to clear gastric obstructions and impaction of material. Some have presented such a level of economic damage to owner's property that they have been euthanased, but most owners of fabric-eating cats seem remarkably tolerant of their cat's behaviour.

The exact cause of the behaviour is unknown but a variety of factors have been suggested, including genetic. Wool eating is suggested to be a heritable trait, although it appears to be triggered by environmental factors, of which the most common is associated with transfer to a new household. It has been suggested that it could be based upon a physiological hyperactivity of the autonomic nervous system. Such neuronal disturbances could affect the control of the digestive tract and thereby produce unusual food cravings and inappropriate appetite stimulation, although the exact mechanism is unclear.

Another suggestion is that the desire to suck and knead wool, and then other fabric, is a continuing redirected form of suckling behaviour, resulting from the failure of the cat to mature fully. Some cats grow out of the behaviour at maturity, but others will eat all unattended fabric despite good nutrition and husbandry. The behaviour is usually more prevalent in cats housed permanently indoors. Fabric eating is sometimes secretive, but it is usually blatant and unaffected by punishment. While most cats will consume fabric at any time, some will take a woollen item to the food bowl and eat this alternately with their usual diet and only at meal times.

Some fabric-eating cats have caused hundreds of pounds worth of damage to designer jumpers, carpets and tweed-covered furniture! Fabric-eating behaviour may sometimes be triggered by some form of stress, perhaps in the form of medical treatment, or the introduction of another cat to the household. A significant number of cats in the survey first exhibited fabric eating within one month of acquisition. Insufficient handling of kittens before adoption or separation from the mother at too early an age may also lead to stress and trigger the behaviour. Only 15% of cats in the survey were acquired before eight weeks of age and earlier than often recommended for maximizing the prospect of as full an emotional development as possible of the later maturing pedigree oriental strains in particular. Over half the pedigree cats in the survey were acquired by their owners at or beyond the minimum age of 12 weeks recommended by the UK Governing Council of the Cat Fancy, suggesting that the age at which a kitten is taken from its mother may not be the only influence of the subsequent development of fabric eating.

Continuing infantile traits, such as over-dependence on the physical presence of the owner, can lead to separation anxiety when the owner departs, and cause the cat to start to eat fabric. The behaviour may then be triggered during subsequent stressful experiences as a learned pattern, even in response to previously tolerated influences. The best hope for treatment of fabric eating probably rests with such cases, where the relationship between owner and cat can be modified so that the cat is made less dependent on its owners for emotional security, and the need to eat fabric as a form of anxiety-relieving displacement behaviour can be reduced.

Fabric eating also seems to form part of a prey catching/ingestion sequence otherwise usually unexpressed in the day-to-day repertoire of the pet cat fed prepared and often very easily digested food. Indeed, 40% of the cats reported in the survey had little or no access to the outdoors and hence restricted or no opportunity to develop exploratory and hunting behaviour, including ingestion of small prey.

Treatment

Treatment currently involves a combination approach of social restructuring with the owner (see 'Over-attachment' above), increasing the level of stimulation for the cat through play, increased activity at home, opportunity to investigate novel stimuli and, where possible, the opportunity for indoor cats to lead an outdoor life. This can mean allowing free access or housing in a secure pen or accustoming the cat to being walked on a lead and harness.

Increasing the fibre content of the diet by offering a dry diet and/or gristly meat attached to large bones to increase food (prey) handling and ingestion time has also brought improvements in many, and even a few total cures. Others have improved by being offered increased fibre in the form of bran or chopped undyed wool or tissue paper blended in with their usual wet canned diet. Such tactics presumably help because the higher fibre intake keeps the cat's stomach active and reduces any appetite-related motivations to fabric eating. Remote ambushes using aromatic taste deterrents such as eucalyptus oil or menthol applied to woollen clothes may be employed, though traditional deterrents using pepper or chilli powder seem only to broaden the cat's normal taste preferences! Remote aversion tactics using touch-sensitive cap exploders under clothes deliberately made available can deter some cats and some may be safely channelled into chewing only certain acceptable items at meal and resting times. Owners of such cats have found that the cat needs to be kept supplied with a cheap supply of its favourite fabric to preserve other clothes and household items. No drugs are currently recognized as being helpful with treatment.

Counselling Owners and Treating Behaviour Problems in Cats

Counselling owners of cats with behavioural problems involves patience, understanding and, most of all, time. Owners will often be distressed not only because of a soiled house or scratched furniture but because they feel that the cat itself is upset. Many problems occur as a result of lack of awareness of the cat's requirements, poor or changed indoor facilities, unrealistic owner expectations or inappropriate interactive behaviour between the owner and cat or cats (Turner, 1991). The initial aim of treatment is therefore to establish communication with the owner, by relaxing and enabling him or her to express their concerns without fear of ridicule or rejection, and to explain why the unusual behaviour may be occurring. This is usually assisted by the presence of the cat itself, even if only in a basket in the surgery. The initial phases of consultation at least should be held in private and the process should not be disturbed. After consultation it is essential that the consultant remains available and contactable by telephone to receive progress reports and deal with continuing concerns.

Very few problems have simple answers, and most necessitate the recording of a complete problem history, relevant medical history, lifestyle and relationship data within the family. The nature of the home environment is crucial, both to the cause of many problems and, through modification of access, in the treatment of most.

Treatment of many cases may be facilitated in the short term by supportive drug therapy, though this can rarely be recommended long term. Progestins (artificial progestagen) are most frequently used on a tapered dose prescription over a few weeks as 'calming agents' to assist behavioural modification techniques rather than for any identifiable hormonal influence. Anti-androgenic injectables are also employed, most frequently on a single dose basis, for the same reasons, and sedatives and other compounds such as antidepressants and morphine antagonists may be used as required in the treatment of obsessive compulsive disorders such as self-mutilation. Occasionally, homoeopathic or Bach's Flower Remedy treatments have also been employed under the guidance of a veterinary surgeon with a special interest in that area. Interestingly, with many cases of aggression or indoor spraying by oriental breeds, the alternative approaches were found to be more helpful than traditional medicine, and although they cannot yet be explained by known science, they should not be overlooked. In all cases, drug support of any kind is rarely curative and usually inhibits learning. The use of drugs often has to be based on a trial and error approach as there are marked variations in response between breeds and individuals. Where used, drugs are offered as short-term vehicles to facilitate acceptance by the cat of management, husbandry or behaviour modification techniques and to facilitate learning by degree during systematic desensitization. Prescription of certain well-marked psychoactive drugs, especially progestins, without any behaviour modification advice has often been tried by veterinary practitioners prior to referral and been found to be ineffective or only effective at modifying the cat's behaviour for the period of dosage. Progestins especially are usually best only offered after a failed period of treatment without drug support and only on a reducing dose for no more than three to four weeks and in support of behaviour modification advice.

After consultation, a letter summarizing the analysis and cause(s) of the problem, suggestions for treatment, rationale behind any accompanying supportive drug therapy, and perhaps likelihood and timescale for successful treatment should be forwarded to the owner.

References

Ablett, P. (1981) Public reaction to control. In: *The Ecology and Control of Feral Cats.* Universities Federation for Animal Welfare, Potters Bar, pp. 60–62.

Adamec, R.E. (1976) The interaction of hunger and preying in the domestic cat (*Felis catus*): an adaptive hierarchy? *Behavioural Biology* 18, 263–272.

Adamec, R.E., Stark-Adamec, C. and Livingstone, K.E. (1980) The development of predatory aggression and defense in the domestic cat (*Felis catus*). *Behavioural and Neural Biology* 30, 389–409.

Adamec, R.E., Stark-Adamec, C. and Livingstone, K.E. (1983) The expression of an early developmentally emergent defensive bias in the adult domestic cat (*Felis catus*) in non-predatory situations. *Applied Animal Ethology* 10, 89–108.

Aldis, O. (1975) *Play fighting.* Academic Press, New York.

Andres, K.H. (1969) Der olfactorische Saum der Katze. *Zeitschrift für Zellforschung und mikroskopische Anatomie* 96, 250–274.

Anon. (1990) Cats across the world. *Anthrozoös* 3, 196.

Association of Pet Behaviour Counsellors (1990) *Annual Report 1990.* APBC, London.

Baerends-van Roon, J.M. and Baerends, G. (1979) *The Morphogenesis of the Behaviour of the Domestic Cat: with a special emphasis on the development of prey-catching.* North Holland, Amsterdam.

Bard, P. and Macht, M.B. (1958) The behaviour of chronically decerebrate cats. In: Wolstenholme, G.E.W. and O'Connor, C.M. (eds), *CIBA Foundation Symposium on the Neurological Basis of Behaviour.* J. & A. Churchill Ltd, London, pp. 55–71.

Barrett, P. and Bateson, P. (1978) The development of play in cats. *Behaviour* 66, 106–120.

Bateson, P., Mendl, M. and Feaver, J. (1990) Play in the domestic cat is enhanced by rationing of the mother during lactation. *Animal Behaviour* 40, 514–525.

Belkin, M., Yinon, U., Rose, L. and Reisert, I. (1977) Effect of visual environment on refractive error of cats. *Documenta Ophthalmologica* 42, 433–437.

Berkley, M.A. (1976) Cat visual psychophysics: neural correlates and comparisons with man. *Progress in Psychobiology and Physiological Psychology* 6, 63–119.

Biben, M. (1979) Predation and predatory play behaviour of domestic cats. *Animal Behaviour* 27, 81–94.

Blakemore, C. and Van Sluyters, R.C. (1975) Innate and environmental factors in the development of the kitten's visual cortex. *Journal of Physiology* 248, 663–716.

Boudreau, J.C. (1989) Neurophysiology and stimulus chemistry of mammalian taste systems. In: Teranishi, R., Buttery, R.G. and Shahidi, F. (eds), *Flavor Chemistry: Trends and Developments. ACS Symposium Series 388*, pp. 122–137.

Boudreau, J.C., Oravec, J. and White, T.D. (1981) A peripheral neural correlate of the human fungiform papillae; sweet, bitter sensations. *Chemical Senses* 6, 129–141.

Bradshaw, J.W.S. (1986) Mere exposure reduces cats' neophobia to unfamiliar food. *Animal Behaviour* 34, 613–614.

Bradshaw, J.W.S. (1991) Sensory and experiential factors in the design of foods for domestic dogs and cats. *Proceedings of the Nutrition Society* 50, 99–106.

Bradshaw, J.W.S. and Nott, H.M.R. (1992) Social and communication behaviour of companion dogs. In: Serpell, J.A. (ed.), *The Domestic Dog: The Biology of its Behaviour.* Cambridge University Press, Cambridge, in press.

Bravo, M., Blake, R. and Morrison, S. (1988) Cats see subjective contours. *Vision Research* 28, 861–865.

Broom, D.M. (1988) The scientific assessment of animal welfare. *Applied Animal Behaviour Science* 20, 5–19.

Brown, K.A., Buchwald, J.S., Johnson, J.R. and Mikolich, D.J. (1978) Vocalization in the cat and kitten. *Developmental Psychobiology* 11, 559–570.

Burger, I.H. (ed.) (1990) *Pets, Benefits and Practice (Waltham Symposium 20).* British Veterinary Association Publications, London.

Burgess, P.R. and Perl, E.R. (1973) Cutaneous mechanoreceptors and nociceptors. In: Iggo, A. (ed.), *Handbook of Sensory Physiology, Vol. II: Somatosensory System.* Springer-Verlag, New York, pp. 29–78.

Caro, T.M. (1980) Effects of the mother, object play, and adult experience on predation in cats. *Behavioural and Neural Biology* 29, 29–51.

Caro, T.M. (1981) Predatory behaviour and social play in kittens. *Behaviour* 76, 1–24.

Carpenter, J.A. (1956) Species differences in taste preferences. *Journal of Comparative and Physiological Psychology* 49, 139–144.

Chesler, P. (1969) Maternal influence in learning by observation in kittens. *Science* 166, 901–903.

Childs, J.E. (1990) Urban cats: their demography, population density, and owner characteristics in Baltimore, Maryland. *Anthrozoös* 3, 234–244.

Chipman, P. (1990) Influence on the Home Range Sizes of Domestic Cats (*Felis catus*), in an Urban Environment. MSc Thesis, Manchester Polytechnic.

Clark, J.M. (1975) The effects of selection and human preference on coat colour gene frequencies in urban cats. *Heredity* 35, 195–210.

Clutton-Brock, J. (1987) *A Natural History of Domesticated Mammals.* Cambridge University Press, Cambridge, and the British Museum (Natural History), London.

Colgan, P. (1989) *Animal Motivation.* Chapman & Hall, London.

Collard, R.R. (1967) Fear of strangers and play behavior in kittens with varied social experience. *Child Development* 38, 877–891.

Cook, N.E., Kane, E., Rogers, Q.R. and Morris, J.G. (1985) Self-selection of dietary casein and soy-protein by the cat. *Physiology and Behavior* 34, 583–594.

Corbett, L.K. (1979) Feeding Ecology and Social Organisation of Wildcats (*Felis silvestris*) and Domestic Cats (*Felis catus*) in Scotland. PhD Thesis, University of Aberdeen.

Costalupes, J.A. (1983) Temporal integration of pure tones in the cat. *Hearing Research* 9, 43–54.

Crouch, J.E. (1969) *Text-Atlas of Cat Anatomy,* Lea and Febiger, Philadelphia.

Dards, J.L. (1983) The behaviour of dockyard cats: interactions of adult males. *Applied Animal Ethology* 10, 133–153.

Davey, G. (1989) *Ecological Learning Theory.* Routledge, London.

Davis, R.G. (1973) Olfactory psychophysical parameters in man, rat, dog, and pigeon. *Journal of Comparative and Physiological Psychology* 85, 221–232.

de Boer, J.N. (1977) Dominance relations in pairs of domestic cats. *Behavioural Processes* 2, 227–242.

Deag, J.M., Manning, A. and Lawrence, C.E. (1988) Factors influencing the mother–kitten relationship. In: Turner, D.C. and Bateson, P. (eds), *The Domestic Cat: the Biology of its Behaviour.* Cambridge University Press, Cambridge, pp. 23–39.

Eccles, R. (1972) Autonomic innervation of the vomeronasal organ of the cat. *Physiology and Behavior* 28, 1011–1015.

Edney, A.T.B. (ed.) (1988) *The Waltham Book of Dog and Cat Nutrition,* 2nd edn. Pergamon, Oxford.

Elul, R. and Marchiafava, P.L. (1964) Accommodation of the eye as related to behaviour in the cat. *Archives Italienne de Biologie* 102, 616–644.

Everett, G.M. (1944) Observations on the behavior and neurophysiology of acute thiamin deficient cats. *American Journal of Physiology* 141, 439–448.

Evinger, C. and Fuchs, A.F. (1978) Saccadic, smooth pursuit and optokinetic eye movements of the trained cat. *Journal of Physiology* 285, 209–229.

Ewer, R.F. (1973) *The Carnivores.* Weidenfield and Nicolson, London.

Fay, R.R. (1988) Comparative psychoacoustics. *Hearing Research* 34, 295–306.

Feaver, J., Mendl, M. and Bateson, P. (1986) A method for rating the individual distinctiveness of domestic cats. *Animal Behaviour* 34, 1016–1025.

Fitzgerald, B.M. (1988) Diet of domestic cats and their impact on prey populations. In: Turner, D.C. and Bateson, P. (eds), *The Domestic Cat: The Biology of its Behaviour.* Cambridge University Press, Cambridge, pp. 123–147.

Fox, M.C. (1974) *Understanding Your Cat.* Coward, McCann and Geoghegan, New York.

Fox, R. and Blake, R. (1971) Stereoscopic vision in the cat. *Nature* 233, 55–56.

Frazer-Sissom, D.E., Rice, D.A. and Peters, G. (1991) How cats purr. *Journal of Zoology, London* 223, 67–78.

Gittleman, J.L. (1991) Carnivore olfactory bulb size: allometry, phylogeny and ecology. *Journal of Zoology, London* 225, 253–272.

Gordon, F. and Jukes, M.G.M. (1964) Dual organisation of the exteroceptive components of the cat's gracile nucleus. *Journal of Physiology* 139, 385–399.

Gorman, M.L. and Trowbridge, B.J. (1989) The role of odor in the social lives of carnivores. In: Gittleman, J.L. (ed.), *Carnivore Behavior, Ecology, and Evolution*. Chapman and Hall, London, pp. 57–88.

Grastyan, E. and Vereczkei, L. (1974) Effects of spatial separation of the conditioned signal from the reinforcement: a demonstration of the conditioned character of the orienting response or the orientational character of conditioning. *Behavioural Biology* 10, 121–146.

Gray, J.A.B. (1966) The representation of information about rapid changes in a population of receptor units signalling mechanical events. In: de Reuck, A.V.S. and Knight, J. (eds), *Touch, Heat and Pain*. CIBA Foundation, London, pp. 299–315.

Guilford, T. and Dawkins, M.S. (1987) Search images not proven: a reappraisal of recent evidence. *Animal Behaviour*, 35, 1838–1845.

Gunter, R. (1951) Visual size constancy in the cat. *British Journal of Psychology* 42, 288–293.

Hart, B.L. (1979) Breed-specific behaviour. *Feline Practice* 9 (6), 10–13.

Hart, B.L. and Barrett, R.E. (1973) Effects of castration on fighting, roaming and urine spraying in adult male cats. *Journal of the American Veterinary Medical Association* 163, 290–292.

Hart, B.L. and Hart, L.A. (1985) *Canine and Feline Behavioural Therapy*. Lea and Febiger, Philadelphia.

Hart, B.L. and Leedy, M.G. (1987) Stimulus and hormonal determinants of Flehmen behaviour in cats. *Hormones and Behaviour* 21, 44–52.

Haskins, R. (1977) Effect of kitten vocalizations on maternal behavior. *Journal of Comparative and Physiological Psychology* 91, 830–838.

Haskins, R. (1979) A casual analysis of kitten vocalization: an observational and experimental study. *Animal Behaviour* 27, 726–736.

Haspel, C. and Calhoon, R.E. (1990) The interdependence of humans and free-ranging cats in Brooklyn, New York. *Anthrozoös* 3, 155–160.

Heffner, R.S. and Heffner, H.E. (1985) Hearing range of the domestic cat. *Hearing Research* 19, 85–88.

Heffner, R.S. and Heffner, H.E. (1988) Sound localization acuity in the cat: effect of azimuth, signal duration, and test procedure. *Hearing Research* 36, 221–232.

Hein, A. and Held, R. (1967) Dissociation of the visual placing response into elicited and guided components. *Science* 158, 390–391.

Hobhouse, L.T. (1915) *Mind in Evolution*, 2nd edn. MacMillan, London.

Houpt, K.J. and Wolski, T.R. (1982) *Domestic Animal Behaviour for Veterinarians and Animal Scientists*. Iowa State University Press, Ames.

Hughes, A. (1972) Vergence in the cat. *Vision Research* 12, 1961–1964.

Hughes, A. (1977) The topography of vision in mammals of contrasting life style: comparative optics and retinal organisation. In: Crescitelli, F. (ed.), *The Visual System in Vertebrates*, Springer-Verlag, Berlin, pp. 613–756.

Iggo, A. (1966) Cutaneous receptors with a high sensitivity to mechanical displacement. In: de Reuck, A.V.S. and Knight, J. (eds), *Touch, Heat and Pain*. CIBA Foundation, London, pp. 237–256.

Iggo, A. (1982) Cutaneous sensory mechanisms. In: Barlow, H.B. and Mollon, J.D. (eds), *The Senses*. Cambridge University Press, Cambridge, pp. 369–408.

Jalowiec, J.E., Panksepp, J., Shabshelowitz, H., Zolovick, A.J., Stern, W. and

Morgane, P.J. (1973) Suppression of feeding in cats following 2-deoxy-D-glucose. *Physiology and Behavior* 10, 805–807.

Jerison, H.J. (1985) Animal intelligence as encephalisation. *Philosophical Transactions of the Royal Society of London B*, 308, 21–35.

John, E.R., Chesler, P., Bartlett, F. and Victor, I. (1968) Observational learning in cats. *Science* 159, 1489–1491.

Johnson, R.F., Randall, S. and Randall, W. (1983) Freerunning and entrained circadian rhythms in activity, eating and drinking in the cat. *Journal of Interdisciplinary Cycle Research* 14, 315–327.

Kalmus, H. (1955) The discrimination by the nose of the dog of individual human odours and in particular of the odours of twins. *British Journal of Animal Behaviour* 3, 25–31.

Kane, E. (1989) Feeding behaviour of the cat. In: Burger, I.H. and Rivers, J.P.W. (eds), *Nutrition of the Dog and Cat*. Cambridge University Press, Cambridge, pp. 147–158.

Karsh, E.B. and Turner, D.C. (1988) The human–cat relationship. In: Turner, D.C. and Bateson, P. (eds), *The Domestic Cat: The Biology of its Behaviour*. Cambridge University Press, Cambridge, pp. 159–177.

Katcher, A.H. and Beck, A.M. (eds) (1983) *New Perspectives on Our Lives with Companion Animals*. University of Pennsylvania Press, Philadelphia.

Kaufman, L.W., Collier, G., Hill, W.L. and Collins, K. (1980) Meal cost and meal patterns in an uncaged domestic cat. *Physiology and Behavior* 25, 135–137.

Kerby, G. (1987) The Social Organisation of Farm Cats (*Felis catis* L.). DPhil Thesis, University of Oxford.

Kerby, G. and Macdonald, D.W. (1988) Cat society and the consequences of colony size. In: Turner, D.C. and Bateson, P. (eds), *The Domestic Cat: The Biology of its Behaviour*. Cambridge University Press, Cambridge, pp. 67–81.

Keverne, E.B., Murphy, C.L., Silver, W.L., Wysocki, C.J. and Meredith, M. (1986) Non-olfactory chemoreceptors of the nose: recent advances in understanding the vomeronasal and trigeminal systems. *Chemical Senses* 11, 119–133.

Kiley-Worthington, M. (1976) The tail movements of ungulates, canids and felids with particular reference to their causation and function as displays. *Behaviour* 56, 69–115.

Kiley-Worthington, M. (1984) Animal language? Vocal communication of some ungulates, canids and felids. *Acta Zoologica Fennica* 171, 83–88.

Kitchener A. (1991) *The Natural History of the Wild Cats*. Christopher Helm, London.

Kolb, B. and Nonneman, A.J. (1975) The development of social responsiveness in kittens. *Animal Behaviour* 23, 368–374.

Kuo, Z.Y. (1960) Studies on the basic factors in animal fighting. VII. Interspecies coexistence in mammals. *Journal of Genetic Psychology* 97, 211–225.

Law, G., Boyle, H., Johnston, J. and Macdonald, A. (1990) Food presentation: Part 2 – Cats. *RATEL* 17, 103–105.

Levine, M.S., Hull, C.D. and Buchwald, N.A. (1980) Development of motor activity in kittens. *Developmental Psychobiology* 13, 357–371.

Levine, M.S., Lloyd, R.L., Fisher, R.S., Hull, C.D. and Buchwald, N.A. (1987) Sensory, motor and cognitive alterations in aged cats. *Neurobiology of Aging* 8, 253–263.

Leyhausen, P. (1979) *Cat Behavior: The Predatory and Social Behavior of Domestic and Wild Cats.* Garland STPM Press, New York.

Leyhausen, P. (1988) The tame and the wild – another Just-So Story? In: Turner, D.C. and Bateson, P. (eds), *The Domestic Cat: The Biology of its Behaviour.* Cambridge University Press, Cambridge, pp. 57–66.

Liberg, O. and Sandell, M. (1988) Spatial organisation and reproductive tactics in the domestic cat and other felids. In: Turner, D.C. and Bateson, P. (eds), *The Domestic Cat: The Biology of its Behaviour.* Cambridge University Press, Cambridge, pp. 83–98.

Loop, M.S. and Frey, T.J. (1981) Critical flicker fusion in Siamese cats. *Experimental Brain Research* 43, 65–68.

Loop, M.S., Millican, C.L. and Thomas, S.R. (1987) Photopic spectral sensitivity of the cat. *Journal of Physiology* 382, 537–553.

Luescher, U.A., McKeown, D.B. and Halip, J. (1991) Stereotypic or obsessive-compulsive disorders in dogs and cats. *Veterinary Clinics of North America: Small Animal Practice* 21, 401–413.

Macdonald, D.W. (1983) The ecology of carnivore social behaviour. *Nature* 301, 379–389.

Macdonald, D.W., Apps, P.J., Carr, G.M. and Kerby, G. (1987) Social dynamics, nursing coalitions and infanticide among farm cats, *Felis catus. Advances in Ethology (supplement to Ethology)* 28, 1–64.

MacDonald, M.L., Rogers, Q.R. and Morris, J.G. (1984) Nutrition of the domestic cat, a mammalian carnivore. *Annual Review of Nutrition* 4, 521–562.

MacDonald, M.L., Rogers, Q.R. and Morris, J.G. (1985) Aversion of the cat to dietary medium-chain triglycerides and caprylic acid. *Physiology and Behavior* 35, 371–375.

MacDonnell, M.F. and Flynn, J.P. (1966) Control of sensory fields by stimulation of hypothalamus, *Science* 152, 1406–1408.

McFarland, D. (1985) *Animal Behaviour: Psychobiology, Ethology and Evolution* Longman, Harlow.

Martin, P. (1984) The (four) whys and wherefores of play in cats: a review of functional, evolutionary, developmental and causal issues. In: Smith, P.K. (ed.), *Play in Animals and Humans.* Basil Blackwell, Oxford, pp. 71–94.

Martin, P. and Bateson, P. (1985) The ontogeny of locomotor play behaviour in the domestic cat. *Animal Behaviour* 33, 502–510.

Martin, P. and Bateson, P. (1988) Behavioural development in the cat. In: Turner, D.C. and Bateson, P. (eds), *The Domestic Cat: The Biology of its Behaviour.* Cambridge University Press, Cambridge, pp. 9–22.

Martin, R.L. and Webster, W.R. (1987) The auditory spatial acuity of the domestic cat in the interaural horizontal and median vertical planes. *Hearing Research* 30, 239–252.

Martin, R.L. and Webster, W.R. (1989) Interaural sound pressure level differences associated with sound-source locations in the frontal hemifield of the domestic cat. *Hearing Research* 38, 289–302.

Mason, G.J. (1991) Stereotypies: a critical review. *Animal Behaviour* 41, 1015–1037.

Mellen, J.D. (1988) The effects of hand-raising on sexual behavior of captive small felids using domestic cats as a model. *Annual Proceedings of the American*

Association of Zoological Parks and Aquariums 1988, 253–259.

Mendl, M. (1988) The effects of litter-size variation on the development of play behaviour in the domestic cat: litters of one and two. *Animal Behaviour* 36, 20–34.

Mendl, M. and Harcourt, R. (1988) Individuality in the domestic cat. In: Turner, D.C. and Bateson, P. (eds), *The Domestic Cat: The Biology of its Behaviour.* Cambridge University Press, Cambridge, pp. 41–54.

Mendoza, D.L. and Ramirez, J.M. (1987) Play in kittens (*Felis domesticus*) and its association with cohesion and aggression. *Bulletin of the Psychonomic Society* 25, 27–30.

Mertens, C. (1991) Human–cat interactions in the human setting. *Anthrozoös* 4, 214–231.

Mertens, C. and Turner, D.C. (1988) Experimental analysis of human–cat interactions during first encounters. *Anthrozoös* 2, 83–97.

Meyer, D.R. and Anderson, R.A. (1965) Colour discrimination in cats. In: de Reuck, A.V.S. and Knight, J. (eds), *Colour Vision: Physiology and Experimental Psychology.* CIBA Foundation, London, pp. 325–344.

Michael, R.P. (1961) Observations upon the sexual behaviour of the domestic cat (*Felis catus* L.) under laboratory conditions. *Behaviour* 18, 1–24.

Miles, R.C. (1958) Learning in kittens with manipulatory, exploratory and food incentives. *Journal of Comparative and Physiological Psychology* 51, 39–42.

Moelk, M. (1944) Vocalizing in the house-cat; a phonetic and functional study. *American Journal of Psychology* 57, 184–205.

Moelk, M. (1979) The development of friendly approach behavior in the cat: a study of kitten–mother relations and the cognitive development of the kitten from birth to eight weeks. In: Rosenblatt, J.S., Hinde, R.A., Beer, C. and Busnel, M. (eds), *Advances in the Study of Behaviour*, Vol. 10. Academic, New York.

Mugford, R.A. (1977) External influences on the feeding of carnivores. In: Kare, M.R. and Maller, O. (eds), *The Chemical Senses and Nutrition.* Academic, New York, pp. 25–50.

Mumma, R. and Warren, J.M. (1968) Two-cue discrimination learning by cats. *Journal of Comparative and Physiological Psychology* 66, 116–121.

National Research Council (1986) *Nutrient Requirements of Cats.* National Academy of Sciences, Washington, DC.

Natoli, E. (1985) Behavioural responses of urban feral cats to different types of urine marks. *Behaviour* 94, 234–243.

Natoli, E. (1990) Mating strategies in cats: a comparison of the role and importance of infanticide in domestic cats, *Felis catus* L., and lions, *Panthera leo* L. *Animal Behaviour* 40, 183–186.

Natoli, E. and de Vito, E. (1991) Agonistic behaviour, dominance rank and copulatory success in a large multi-male feral cat, *Felis catus* L., colony in central Rome. *Animal Behaviour* 42, 227–241.

Neville, P.F. (1990) *Do Cats Need Shrinks?* Sidgewick and Jackson, London.

Neville, P.F. (1991) Treatment of behaviour problems in cats. *Practice* 13, 43–50.

Neville, P.F. and Bradshaw, J.W.S. (1991) Unusual appetites. *Bulletin of the Feline Advisory Bureau* 28, 5–6, 32.

Norton, T.T. (1974) Receptive-field properties of superior colliculus cells and development of visual behaviour in kittens. *Journal of Neurophysiology* 37, 674–690.

Olmstead, C.E. and Villablanca, J.R. (1980) Development of behavioural audition in the kitten. *Physiology and Behaviour* 24, 705–712.

Oswald, I. (1962) *Sleeping and Waking.* Elsevier, Amsterdam.

Packer, C., Gilbert, D.A., Pusey, A.E. and O'Brien S.J. (1991) A molecular genetic analysis of kinship and co-operation in African lions. *Nature* 351, 562–565.

Palen, G.F. and Goddard, G.V. (1966) Catnip and oestrous behaviour in the cat. *Animal Behaviour* 14, 372–377.

Passanisi, W.C. and Macdonald, D.W. (1990a) Group discrimination on the basis of urine in a farm cat colony. In: Macdonald, D.W., Muller-Schwarze, D. and Natynczuk, S.E. (eds), *Chemical Signals in Vertebrates 5.* Oxford University Press, Oxford, pp. 336–345.

Passanisi, W.C. and Macdonald, D.W. (1990b) *The Fate of Controlled Feral Cat Colonies.* Universities Federation for Animal Welfare, Potters Bar.

Pasternak, T. and Merigan, W.H. (1980) Movement detection by cats: invariance with direction and target configuration. *Journal of Comparative and Physiological Psychology* 94, 943–952.

Peters, G. and Wozencraft, W.C. (1989) Acoustic communication by fissiped carnivores. In: Gittleman, J.L. (ed.), *Carnivore Behavior, Ecology, and Evolution.* Chapman and Hall, London, pp. 14–56.

Poucet, B. (1985) Spatial behaviour of cats in cue-controlled environments. *Quarterly Journal of Experimental Psychology* 37B, 155–179.

Radinsky, L. (1975) Evolution of the felid brain. *Brain Behaviour and Evolution,* 11, 214–254.

Randall, W. and Parsons, V. (1987) Three views of the annual phase map of the domestic cat, *Felis catus* L. *Journal of Interdisciplinary Cycle Research* 18, 17–28.

Randi, E. and Ragni, B. (1991) Genetic variability and biochemical systematics of domestic and wild cat populations (*Felis silvestris*: Felidae). *Journal of Mammalogy* 72, 79–88.

Romand, R. and Ehret, G. (1984) Development of sound production in normal, isolated, and deafened kittens during the first postnatal months. *Developmental Psychobiology* 17, 629–649.

Rosenblatt, J.S. (1972) Learning in newborn kittens. *Scientific American* 227, 18–25.

Rowan, A.N. (1988) Animal anxiety and animal suffering. *Applied Animal Behaviour Science* 20, 135–142.

Rowan, A.N. and Williams, J. (1987) The success of companion animal management programs: a review. *Anthrozoös* 1, 110–122.

Rozin, P. (1976) The selection of foods by rats, humans, and other animals. *Advances in the Study of Behaviour* 6, 21–76.

Serpell, J.A. (1986) *In the Company of Animals: a Study of Human–Animal Relationships.* Basil Blackwell, Oxford.

Serpell, J.A. (1988) The domestication and history of the cat. In: Turner, D.C. and Bateson, P. (eds), *The Domestic Cat: The Biology of its Behaviour.* Cambridge University Press, Cambridge, pp. 151–158.

Shettleworth, S.J. (1984) Natural history and evolution of learning in nonhuman mammals. In: Marler, P. and Terrace, H.S. (eds), *The Biology of Learning.* Springer-Verlag, Berlin, pp. 419–433.

Siegel, A. and Pott, C.B. (1988) Neural substrates of aggression and flight in the cat. *Progress in Neurobiology* 31, 261–283.

Smithers, R.H.N. (1968) Cat of the Pharaohs. *Animal Kingdom* 61, 16–23.

Sokolov, E.N. (1963) Higher nervous functions: the orienting reflex. *Annual Review of Physiology* 25, 545–580.

Sparkes, A.H., Adams, D.T., Douthwaite, J.A. and Gruffydd-Jones, T.J. (1990) Assessment of adrenal function in cats: response to intravenous synthetic ACTH. *Journal of Small Animal Practice* 31, 1–5.

Stein, B.E., Magalhâes-Castro, B. and Kruger, L. (1976) Relationship between visual and tactile representations in cat superior colliculus. *Journal of Neurophysiology* 39, 401–419.

Tabor, R. (1983) *The Wildlife of the Domestic Cat.* Arrow Books, London.

Thorne, C.J. (1982) Feeding behaviour in the cat – recent advances. *Journal of Small Animal Practice* 23, 555–562.

Todd, N.B. (1977) Cats and commerce. *Scientific American* 237(5), 100–107.

Triana, E. and Pasnak, R. (1981) Object permanence in cats and dogs. *Animal Learning & Behaviour* 9, 135–139.

Turner, D.C. (1988) Cat behaviour and the human/cat relationship. *Animalis Familiaris* 3, 16–21.

Turner, D.C. (1991) The ethology of the human-cat relationship. *Swiss Archive for Veterinary Medicine* 133, 63–70.

Turner, D.C., Feaver, J., Mendl, M. and Bateson, P. (1986) Variation in domestic cat behaviour towards humans: a paternal effect. *Animal Behaviour* 34, 1890–1901.

Turner, D.C. and Bateson, P. (eds), (1988) *The Domestic Cat: The Biology of its Behaviour.* Cambridge University Press, Cambridge.

Turner, D.C. and Meister, O. (1988) Hunting behaviour of the domestic cat. In: Turner, D.C. and Bateson, P. (eds), *The Domestic Cat: The Biology of its Behaviour.* Cambridge University Press, Cambridge, pp. 111–121.

Turner, D.C. and Stammbach-Geering, M.K. (1990) Owner assessment and the ethology of human–cat relationships. In: Burger, I.H. (ed.), *Pets, Benefits and Practice.* British Veterinary Association Publications, London.

Universities Federation for Animal Welfare (1989) *Feral Cats: Suggestions for Control.* 2nd edn. UFAW, Potters Bar.

Verberne, G. and De Boer, J.N. (1976) Chemocommunication among domestic cats mediated by the olfactory and vomeronasal senses. *Zeitschrift fur Tierpsychologie* 42, 86–109.

Villablanca, J.R. and Olmstead, C.E. (1979) Neurological development of kittens. *Developmental Psychobiology* 12, 101–127.

Voith, V.L. (1981) You, too, can teach a cat tricks (examples of shaping, second-order reinforcement, and constraints on learning). *Modern Veterinary Practice* 62, 639–642.

Warren, J.M. (1960) Oddity learning set in a cat. *Journal of Comparative and Physiological Psychology* 53, 433–434.

Warren, J.M. (1972) Transfer of responses to open and closed shapes in discrimination learning by cats. *Perception & Psychophysics* 12, 449–452.

Warren, J.M. (1976) Irrelevant cues and shape discrimination learning by cats. *Animal Learning and Behaviour* 4, 22–24.

Warren, J.M. and Beck, C.H. (1966) Visual probability learning by cats. *Journal of Comparative and Physiological Psychology* 61, 316–318.

Watt, D.G.D. (1976) Responses of cats to sudden falls: an otolith originating reflex assisting landing. *Journal of Neurophysiology* 39, 257–265.

Wayne, R.K., Benveniste, R.E., Janczewski, D.N. and O'Brien, S.J. (1989) Molecular and biochemical evolution of the Carnivora. In: Gittleman, J.L. (ed.), *Carnivore Behavior, Ecology and Evolution.* Chapman & Hall, London, pp. 465–494.

West, M.J. (1974) Social play in the domestic cat. *American Zoologist* 14, 427–436.

West, M.J. (1979) Play in domestic kittens. In: Cairns, R.B. (ed.), *The Analysis of Social Interactions.* Hillside, New Jersey.

Wetzel, M.C. and Stuart, D.G. (1976) Ensemble characteristics of cat locomotion and its neural control. *Progress in Neurobiology* 7, 1–98.

Wetzel, M.C., Anderson, R.C., Brady, T.H. and Norgren, K.S. (1977) Kinematics of treadmill galloping by cats. III. Coordination during gait conversions and implications for neural control. *Behavioral Biology* 21, 107–127.

White, T.D. and Boudreau, J.C. (1975) Taste preferences of the cat for neurophysiology active compounds. *Physiological Psychology* 3, 405–410.

Wilkinson, F. (1986) Visual texture segmentation in cats. *Behavioural Brain Research* 19, 71–82.

Wilson, V.J. and Melville Jones, G. (1979) *Mammalian Vestibular Physiology.* Plenum Press, New York.

Winslow, C.N. (1938) Observations of dominance-subordination in cats. *Journal of Genetic Psychology* 52, 425–428.

Wolski, D.V.M. (1982) Social behavior of the cat. *Veterinary Clinics of North America: Small Animal Practice* 12, 693–706.

Woodhouse, J.M. and Barlow, H.B. (1982) Spatial and temporal resolution and analysis. In: Barlow, H.B. and Mollon, J.D. (eds), *The Senses.* Cambridge University Press, Cambridge, pp. 133–164.

Index

Abyssinian cat 9, 188
Acinonyx jubatus *see* Cheetah
affiliative behaviour 103, 153, 157,
    162, 185
aggressive behaviour 143, 159,
    174–175, 185, 198–200
    communication 97, 99–101, 103,
      157
agonistic behaviour 47, 86, 146, 149,
    153
    see also aggressive behaviour;
      defensive behaviour
allogrooming 154
allorubbing *see* communication,
    olfactory
anorexia 126
anxiety 178

balance 16–18, 47–50
brain 51–54
    see also hypothalamus
bunting *see* communication, olfactory,
    object-rubbing
Burmese cat 10, 188–189, 197, 200

carpal hairs 20
catnip 95, 110
cheetah 11, 15, 141

Christianity 10
claws 14, 20
    sharpening 109, 189
coat colours 3–6
communication
    constraints on 92–94
    olfactory 103–110, 146
      allorubbing 109, 110, 155–159,
        165–166, 189
      faeces 108
      object rubbing 109–110, 153, 166,
        189
      scratching 108–109, 189
      urine 42, 104–108, 153–154
      see also gape
    social 152–159
    sound 69, 74–75, 93–99, 165
      see also miaows; purr
    visual 99–103
      facial signals 100–101, 102
      tail signals 101–103, 157–158
      whole-body signals 99–100, 101
    with man 165–166

defensive behaviour 74
    communication 97, 99–101, 103,
      157
DNA fingerprinting 150–151
dog, domestic 40–41, 53, 56, 118, 119,

215